AF368915

Lithium-Ion Batteries: Latest Advances, Challenges and Prospects

Lithium-Ion Batteries: Latest Advances, Challenges and Prospects

Editor

Siamak Farhad

MDPI • Basel • Beijing • Wuhan • Barcelona • Belgrade • Manchester • Tokyo • Cluj • Tianjin

Editor
Siamak Farhad
University of Akron
Ohio
USA

Editorial Office
MDPI
St. Alban-Anlage 66
4052 Basel, Switzerland

This is a reprint of articles from the Special Issue published online in the open access journal *Energies* (ISSN 1996-1073) (available at: https://www.mdpi.com/journal/energies/special_issues/LIBLACP).

For citation purposes, cite each article independently as indicated on the article page online and as indicated below:

LastName, A.A.; LastName, B.B.; LastName, C.C. Article Title. *Journal Name* **Year**, *Volume Number*, Page Range.

ISBN 978-3-0365-8274-0 (Hbk)
ISBN 978-3-0365-8275-7 (PDF)

Contents

About the Editor

Siamak Farhad

Dr. Siamak Farhad (Professor) joined the Mechanical Engineering Department of The University of Akron in 2013. He obtained his Ph.D. degree from the University of Waterloo in Canada in 2012, during which he also spent two years as a visiting researcher at the National Research Council (NRC) Canada. Following his doctoral studies, Dr. Farhad worked as a Post-doctoral Fellow in the "Applied Nano-materials & Clean Energy Lab" at the University of Waterloo. During this time, he actively contributed to the GM Electric Vehicle project in 2012 and 2013. With an extensive research background, Dr. Farhad has made significant contributions to the scientific community. He has authored over 100 research articles published in esteemed peer-reviewed journals and presented at major international conferences. His exceptional work has been recognized with various awards, including the Breakthrough Research Project from the National Science Foundation (NSF), Best Paper Award from the ASME Energy Sustainability Conference, Industrial Fellowship award from the Natural Sciences and Engineering Research Council of Canada (NSERC), and the Toyota Motor Manufacturing Canada (TMMC) award. Dr. Farhad currently serves as the Director of the Center for Precision Manufacturing, which focuses on both research and workforce development initiatives. He has been successful in securing multiple external grants and funds from prestigious organizations such as the NSF, State of Ohio, Department of Defense (DOD), and industry partners. With a diverse professional background, Dr. Farhad brings a wealth of experience, having worked for 10 years in academia, 4 years in the industry, and 4 years in research centers. Through his research, publications, awards, and leadership roles, Dr. Siamak Farhad has established himself as a highly respected expert in the field of mechanical engineering, making significant contributions to clean energy technologies.

Preface to "Lithium-Ion Batteries: Latest Advances, Challenges and Prospects"

This book is dedicated to exploring fundamental and applied research on lithium-ion batteries, with the primary goal of presenting state-of-the-art knowledge and cutting-edge technology advancements. It provides extensive coverage of various research aspects, including the development and investigation of new energy materials, advanced modeling techniques, and rigorous testing methodologies. This book also delves into the emerging field of all-solid-state lithium batteries, which have garnered significant attention for their potential to revolutionize the energy storage landscape by offering enhanced safety, higher energy density, and a prolonged cycle life. Furthermore, this book emphasizes the critical importance of battery recycling and remanufacturing practices. It addresses the growing concerns surrounding sustainable battery technologies and their environmental impact. By discussing strategies and methodologies for efficient battery recycling and the remanufacturing of battery components, this book aims to contribute to the development of environmentally conscious practices in the industry.

Overall, this comprehensive resource serves as a valuable reference for researchers, engineers, and individuals interested in gaining an understanding of the latest advancements, emerging trends, and sustainability aspects related to lithium-ion batteries.

Siamak Farhad
Editor

Article

Performance of Cathodes Fabricated from Mixture of Active Materials Obtained from Recycled Lithium-Ion Batteries

Hammad Al-Shammari [1,2] and Siamak Farhad [1,*]

1 Advanced Energy & Manufacturing Laboratory, Department of Mechanical Engineering, University of Akron, Akron, OH 44325, USA; hmtshammari@ju.edu.sa
2 Department of Mechanical Engineering, Jouf University, Sakaka 72388, Saudi Arabia
* Correspondence: sfarhad@uakron.edu

Abstract: The cathode performance of lithium-ion batteries (LIBs) fabricated from recycled cathode active materials is studied for three scenarios. These scenarios are based on the conditions for separation of different cathode active materials in recycling facilities during the LIB's recycling process. In scenario one, the separation process is performed ideally, and the obtained pure single cathode active material is used to make new LIBs after regeneration. In scenario two, the separation of active materials is performed with efficiencies of less than 100%, which is the actual case in the recycling process. In this scenario, a single cathode active material that contains a little of the other types of cathode active materials is used to make new LIBs after the materials' regeneration. In scenario three, the separation has not been performed during the recycling process. In this scenario, all types of cathode active materials are regenerated together, and a mixture is used to make new LIBs. The studies are performed through modeling and computer simulation, and several experiments are conducted for validation purposes. The cathode active materials that are studied are the five commercially available cathodes made of $LiMn_2O_4$ (LMO), $LiCoO_2$ (LCO), $LiNi_xMn_yCo_{(1-x-y)}O_2$ (NMC), $LiNi_xCo_yAl_{(1-x-y)}O_2$ (NCA), and $LiFePO_4$ (LFP). The results indicate that the fabrication of new LIBs with a mixture of cathode active materials is possible when cathode active materials are not ideally separated from each other. However, it is recommended that the separation process is added to the recycling process, at least for the separation of LFP or reducing its amount in the cathode active materials mixture. This is because of the difference of the voltage level of LFP compared to the other studied active materials for cathodes.

Keywords: lithium-ion battery; recycling; cathode performance; mixture of cathode active materials; separation of cathode active materials

Citation: Al-Shammari, H.; Farhad, S. Performance of Cathodes Fabricated from Mixture of Active Materials Obtained from Recycled Lithium-Ion Batteries. *Energies* **2022**, *15*, 410. https://doi.org/10.3390/en15020410

Academic Editor: Carlos Miguel Costa

Received: 3 December 2021
Accepted: 1 January 2022
Published: 6 January 2022

Publisher's Note: MDPI stays neutral with regard to jurisdictional claims in published maps and institutional affiliations.

1. Introduction

Lithium-ion batteries (LIBs) are storage systems for electrical energy. Their relatively high energy density, high power density, and long lifespan have led to the industry making them the first candidate for portable electronics, electric vehicles, and even renewable energy storage [1,2]. A commercial LIB consists of six components: (1) an anode or negative electrode; (2) a cathode or positive electrode; (3) an electrolyte; (4) a separator; (5) current collectors for positive and negative electrodes, which are usually aluminum and copper foils, respectively, and (6) the battery casing, which is usually stainless steel for cylindrical LIBs and polymer coated aluminum for pouch/prismatic LIBs [3]. The negative and positive electrodes are usually made of three materials: the active material for storage of lithium, the conductive material to enhance the electron conductivity of the electrode, and the binder to bond the active and conductive materials and to adhere the electrode to the current collector. Among the anode active materials, graphite is currently the most common one because of its relatively high energy density (372 mAh/g), good mechanical and chemical stabilities, and low cost [4]. Another anode active material is $Li_4Ti_5O_{12}$

(LTO), which is one of the promising anode active materials for LIBs because of its constant charge/discharge profile at 1.5 V versus lithium and its excellent Li-insertion/extraction reversibility with no structural change [5]. For cathode active materials, $LiCoO_2$ (LCO) is widely utilized due to its high specific energy [6]. Despite its success in the LIBs industry, it has some shortcomings such as the high cost (cobalt is not an abundant element on earth) and relatively lower specific power. These shortcomings have led the industry to adopt other materials such as $LiFePO_4$ (LFP), $LiMn_2O_4$ (LMO), $LiNi_xCo_yAl_{(1-x-y)}O_2$ (NCA), and $LiNi_xMn_yCo_{(1-x-y)}O_2$ (NMC) [7]. Each of these cathode active materials has its own advantages and disadvantages. The electrolyte in LIBs is responsible for transferring lithium ions between the cathode and anode active materials. Most commercial LIBs currently use organic liquid electrolytes because of their relatively wide electrochemical stability. An organic liquid electrolyte is comprised of a lithium salt, such as lithium hexafluorophosphate ($LiPF_6$), dissolved in an organic solvent [8]. The separator is an electron non-conductive material that separates the positive electrode from the negative electrode to prevent the short circuit between them [9], while allowing rapid transport of ionic charges between the negative and positive electrodes. In most LIBs, separators are currently made of either microporous polymeric films or nonwoven fabrics. The microporous polymeric films (e.g., polyethylene (PP)) are the preferred separator for LIBs because of their thermal and mechanical stabilities [10].

Among all components/materials of LIBs, the cathode active material is usually expensive, as it may be valued at even 40% of the total material cost of LIBs [11]. Improvement of the energy density of active materials has been a pursuit of the scientific community for a long time [12]. Many researchers have studied the combination/mixture of two or three active materials to improve the electrode performance and reduce the cost of the cathode. For example, in a study, the authors have investigated the cathode mixture of LMO and NCA [13]. In fact, NCA shows a high energy density and a good lifetime, but it shows poor thermal stability at elevated temperatures. On the other hand, LMO shows a better thermal stability, higher nominal voltage, higher power density, and lower cost, but it has a lower energy density. Therefore, the performance of the cathode can be engineered when NCA and LMO are blended, and the cathode is made from a mixture of two active materials. The results of this research showed that, at low C-rates, NCA showed higher specific energy, while at high C-rates LMO showed higher specific energy. Another research has been conducted to investigate the blended electrode of two cathode active materials of LMO and NMC [14]. The results indicated that the low capacity of LMO is increased when it is blended with NMC. A research team has conducted a study on the ternary blend of NMC, LMO, and LMFP and showed the advantage of this blend [15]. Out of the many blended seniors, they showed that the blend of 75% NMC, 12.5% LMFP, and 12.5% LMO is comparable with NMC.

All prior blending studies have been conducted to either reduce the cost or enhance the electrochemical performance of the cathode. In contrast to prior studies, our focus in this paper is on the blended cathode materials that are obtained from the recycling of LIBs using the physical recycling method.

All LIBs that are produced today will be retired after about 10 years. If LIBs are landfilled, we may expect several environmental problems: (1) Contamination: chemicals found in batteries may leak from the casing once the battery is in a landfill and contaminate the area and groundwater. This is a serious threat to the environment, ecosystems, and human health. (2) Safety: the landfilled lithium-ion batteries may catch fire and an explosion may happen. (3) Sustainability: several rare materials such as lithium and cobalt are wasted in landfills and new materials will need to be extracted from mines. In general, three methods are available to recycle LIBs. Hydrometallurgy (melting), pyrometallurgy (chemicals), and physical or direct methods. The direct method is the most cost-effective and environmentally friendly method for recycling LIBs, and it is also the most promising recycling method for these batteries. One advantage of the physical method is that the electrode active materials can be separated without changing the morphology of the materials. Hence, the recycled electrode materials can be regenerated and reused to make new LIBs. In recycling plants that operate based on the physical method, all types of LIBs are usually recycled together without sorting them out based on the battery chemistry. In fact, sorting LIBs is not usually logistically possible. Therefore, the obtained cathode active materials are a blend of different active materials. This motivated us to look at the blend of different cathode materials from the angle of the obtained materials from LIB recycling with the physical method.

In this study, we decide to show how the battery performance can be affected if it is built with a blend of recycled cathode active materials. The focus will be on the electrochemical performance of the blended/mixture of the five most commercially available cathode types: LFP, LMO, NMC, NCA, and LCO. This study is the continuation of the studies conducted by the authors for recycling LIBs [16–19]. The authors' results have already indicated that the complete separation of the five cathode active materials is achievable. However, there is still a concern or uncertainty about the performance of separation that cannot reach 100% in practice. Therefore, we study the performance of LIBs made from the recycled cathode materials based on the following three scenarios:

(I)   The separation process is performed ideally. In this scenario, a pure single regenerated cathode active material is used to make new LIBs.

(II) The separation process is performed with an efficiency of less than 100%. This is the actual scenario in the recycling process with the physical method. In this scenario, a single regenerated cathode active material that contains a little of the other types of cathode active materials is used to make new LIBs.

(III) The separation has not been performed. In this scenario, all types of cathode active materials are regenerated together and used to make new batteries.

In this paper, the results of both mathematical modeling and experiments are presented. For the modeling, a pseudo-two-dimensional (P2D) model based on the porous theory proposed by Newman [20] has been adopted. The computer simulation has been carried out in the COMSOL Multiphysics software package for all three scenarios. The experiments have been conducted to validate the simulation results.

2. Model

Since the objective is to study the performance of the cathode active material, we modeled a half-cell of the LIB. As shown in Figure 1, a half-cell consists of a lithium foil anode, separator, cathode (positive electrode), aluminum correct collector, and liquid electrolyte, which is available in the separator and the cathode.

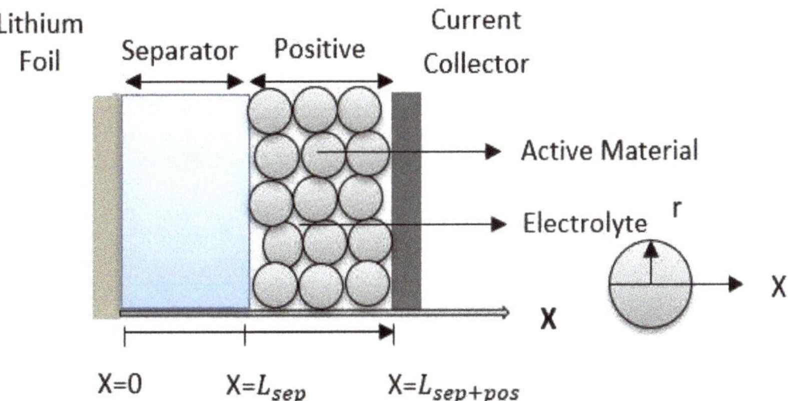

Figure 1. Schematic of the half-cell lithium-ion battery made from a single cathode active material.

The electrochemical reaction during charging and discharging of the half-cell can be expressed as Equation (1) [21]. During discharging, the lithium ions move from the lithium foil to the cathode particles. During charging, the lithium ions move back from the cathode to the lithium foil.

$$xLi^+ + xe^- + \mathcal{M} \underset{\underset{discharge}{\Rightarrow}}{\overset{\overset{charge}{\Leftarrow}}{\rule{3cm}{0pt}}} Li_x\mathcal{M} \tag{1}$$

2.1. Half-Cell Made from a Single Cathode Active Material

The model considers one active material for the cathode, as seen in Figure 1. The negative electrode in the half-cell works as a reference electrode for the battery. For the modeling, we consider the: (a) electric charge transfer in the cathode; (b) ionic charge transfer in the cathode; (c) lithium-ion mass transfer in the cathode; (d) lithium mass transfer in the cathode active material particles (intercalation); (e) ionic charge transfer in the electrolyte; (f) lithium-ion mass transfer in the separator, and (g) electrochemical reaction at the active sites of the cathode. The modeling equations and their boundary and initial conditions are summarized in Table 1. Several assumptions have been made for the modeling. The main assumptions are that the battery is fresh and there is no sign of materials degradation, the cathode active materials are solid spheres with uniform size, the Bruggeman assumption is valid for calculation of the effective conductivities and diffusivities, there is no volume change in the cathode active materials during charging and discharging, no SEI layer is formed on the lithium foil, the entire surface of all cathode active material particles is an active site for electrochemical reactions, the voltage drop in the aluminum current collector is negligible, and the half-cell temperature is kept constant during charging and discharging. In addition, the half-cell model is pseudo-two-dimensional (1D + 1D). This means that the lithium mass transfer equation in the cathode active material particles is solved only in r direction, and all other transport equations are solved in only x direction. See Figure 1 for the r and x directions.

2.2. Half-Cell Made from a Mixture of Cathode Active Materials

The modeling of the cathode performance made from a mixture of cathode active materials (see Figure 2) is similar to the modeling of the cathode made from a single cathode active material. However, there are some differences. The differences are discussed here.

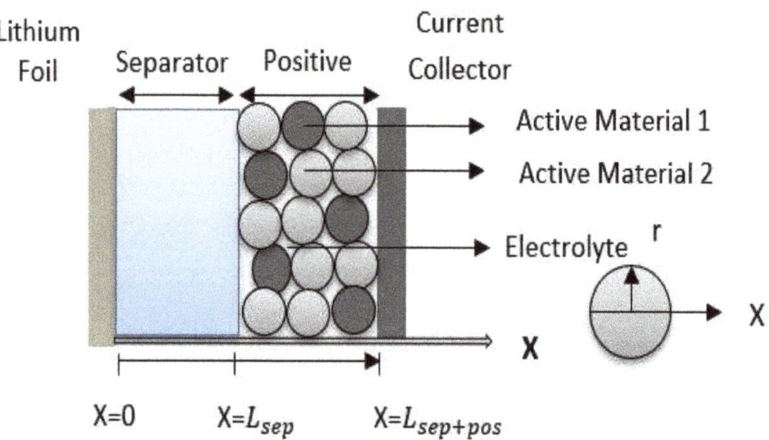

Figure 2. Schematic of the half-cell lithium-ion battery made from a mixture of cathode active materials.

Table 1. The modeling equations for performance simulation of a cathode made from a single active material [20].

Equations	Initial and Boundary Conditions
Li mass transfer in cathode active material particles	
$\frac{\partial c_s}{\partial t} = \frac{1}{r^2}\frac{\partial}{\partial r}\left(r^2 D_s \frac{\partial c_s}{\partial r}\right)$ Solution of this equation gives the distribution of lithium in r direction in the cathode particle, which is located at any x, at any time t	$\frac{\partial c_s}{\partial r} = 0 \text{ at } r = 0$ $\frac{\partial c_s}{\partial r} = -\frac{i_{local}}{F D_s} \text{ at } r = r_p$ $c_s = c_{s,0} \text{ at } t = 0$
Electric charge transfer in the cathode	
$\nabla.i_s = -S_{a,cat}i_{local}$ $i_s = -\sigma_{s,cat,eff}\nabla\phi_s$ $\sigma_{s,cat,eff} = \sigma_{s,cat}\varepsilon_{s,cat}^{1.5}$ $S_{a,cat} = \frac{3}{r_{p,cat}}$ Solutions of these equations give the distribution of electric current along the cathode thickness at any time t	$i_s = 0 \text{ at } x = L_{sep}$ $\phi_s = 0 \text{ at } x = L_{sep}$ $i_s = \frac{I_{applied}}{A} \text{ at } x = L_{sep} + L_{pos}$
Ionic charge transfer in the electrolyte within the cathode	
$\nabla.i_l = S_a i_{local}$ $i_l = -\sigma_{l,cat,eff}\nabla\phi_l + \frac{2RT}{F}\sigma_{l,cat,eff}\left(1 - t_+^0\right)\left(1 + \frac{d\ln f}{d\ln c_l}\right)\nabla\ln c_l$ $\sigma_{l,cat,eff} = \sigma_l\varepsilon_{l,cat}^{1.5}$ Solutions of these equations give the distribution of ionic current along the cathode thickness at any time t	$i_l = 0 \text{ at } x = L_{sep} + L_{pos}$ $i_l = \frac{I_{applied}}{A} \text{ at } x = L_{sep}$
Li$^+$ mass transfer in the electrolyte within the cathode	
$\varepsilon_{l,cat}\frac{\partial c_l}{\partial t} + \nabla.N_l = 0$ $N_l = -D_{l,cat,eff}\nabla c_l - \frac{i_l t_+^0}{F}$ $D_{l,cat,eff} = D_l\varepsilon_{l,cat}^{1.5}$ Solutions of these equations give the distribution of lithium-ion concentration along the cathode thickness at any time t	$N_l = 0 \text{ at } x = L_{sep} + L_{pos}$ $c_l = c_{l,0} \text{ at } t = 0$
Ionic charge transfer in the electrolyte within the separator	
$\nabla.i_l = 0$ $i_l = -\sigma_{l,sep,eff}\nabla\phi_l + \frac{2RT}{F}\sigma_{l,sep,eff}\left(1 - t_+^0\right)\left(1 + \frac{d\ln f}{d\ln c_l}\right)\nabla\ln c_l$ $\sigma_{l,sep,eff} = \sigma_l\varepsilon_{l,sep}^{1.5}$ Solutions of these equations give the distribution of ionic current along the separator thickness at any time t	$i_l = \frac{I_{applied}}{A} \text{ at } x = 0$
Li$^+$ mass transfer in the electrolyte within the separator	
$\varepsilon_{l,sep}\frac{\partial c_l}{\partial t} + \nabla.N_l = 0$ $N_l = -D_{l,sep,eff}\nabla c_l - \frac{i_l t_+^0}{F}$ $D_{l,sep,eff} = D_l\varepsilon_{l,sep}^{1.5}$ Solutions of these equations give the distribution of lithium-ion concentration along the separator thickness at any time t	$N_l = 0 \text{ at } x = 0$ $c_l = c_{l,0} \text{ at } t = 0$
Electrochemical reaction in the cathode (Butler–Volmer kinetics)	
$i_{local,cat} = i_{0,cat}\left[exp\left(\frac{F}{2RT}\eta_{cat}\right) - exp\left(-\frac{F}{2RT}\eta_{cat}\right)\right]$ $i_{0,cat} = Fk_{cat}\left(c_{s,max} - c_{s,surf}\right)^{0.5}c_{s,surf}^{0.5}\left(\frac{c_l}{c_{l,ref}}\right)^{0.5}$ $\eta_{cat} = \phi_s - \phi_l - E_{eq,cat}$ Solutions of these equations give the local current generation along the cathode thickness at any time t as well as the local activation + concentration polarizations along the cathode thickness at any time t	

The mixture of cathode active materials consists of spherical particles from different active materials that are mixed homogenously before the cathode slurry is coated on the aluminum current collector. For the modeling, it is assumed that the mixture is homogeneous, while each type of cathode active material can have its own particle sizes. For the mixture of cathode active materials, the equation of lithium mass transfer should be solved for each cathode active material. The electrochemical reaction equation should also be solved separately for each cathode active material, and the total local current should be obtained from the summation of the local current generated by each cathode active material. The additional modeling equations and their boundary and initial conditions to simulate the behavior of a mixture of cathode active materials are summarized in Table 2.

Table 2. The additional modeling equations required for performance simulation of a cathode made from a mixture of active materials [13].

Equations	Initial and Boundary Conditions
Li mass transfer in the particles of each cathode active material	
$\frac{\partial c_{s,i}}{\partial t} = \frac{1}{r^2}\frac{\partial}{\partial r}\left(r^2 D_{s,i}\frac{\partial c_{s,i}}{\partial r}\right)$ for i^{th} cathode active material Solution of this equation gives the distribution of lithium in r direction in the particles of the cathode active material i, which is located at any x, at any time t	$\frac{\partial c_{s,i}}{\partial r} = 0$ at $r = 0$ $\frac{\partial c_{s,i}}{\partial r} = -\frac{i_{local,i}}{FD_{s,i}}$ at $r = r_{p,i}$ $c_{s,i} = c_{s,i,0}$ at $t = 0$
Electrochemical reaction in the cathode (Butler–Volmer kinetics)	
$i_{local,cat,i} = i_{0,cat,i}\left[exp\left(\frac{F}{2RT}\eta_{cat,i}\right) - exp\left(-\frac{F}{2RT}\eta_{cat,i}\right)\right]$ for ith cathode active material $i_{0,cat,i} = Fk_{cat,i}\left(c_{s,i,max} - c_{s,i,surf}\right)^{0.5} c_{s,i,surf}^{0.5}\left(\frac{C_l}{C_{l,ref}}\right)^{0.5}$ $\eta_{cat,i} = \phi_s - \phi_l - E_{eq,cat,i}$ $i_{loc,total} = \sum_{i=1}^{N} i_{loc,i}$ Solutions of these equations give the total local current generation along the cathode thickness at any time t as well as the local activation + concentration polarizations for each cathode active material along the cathode thickness at any time t	

The capacity of the cathode made from a mixture of cathode active materials can be obtained from Equation (2):

$$Q_{cat} = F\varepsilon_{s,cat}L_{pos}\sum_{i=1}^{N} Y_i c_{s,i,max}\left(SOC_{i,max} - SOC_{i,min}\right) \tag{2}$$

where the state-of-charge (SOC) of each active material can be obtained from Equation (3):

$$SOC_i = \frac{c_{s,i,surf}}{c_{s,i,max}} \tag{3}$$

Based on this mathematical modeling, the simulation of half-cell LIBs was carried out in the COMSOL Multiphysics software package, version 5.2a.

3. Experiment

3.1. Cathode Groups

The cathodes for the experiments are divided into three groups. The first group is the cathodes that were made of a single cathode active material to simulate the performance of the battery in case the separation of active materials during the recycling process is ideal. This group is for the study of scenario I. The second group is the cathodes that were made of a mixture of active materials with one dominant cathode active material to simulate the conditions that the separation process is not ideal and there are some impurities from other types of cathode active materials. This group is for the study of scenario II. The cathodes in this group are made from a dominant cathode active material and the minor percentage of the other four types of cathode active materials that are equally mixed. The third group is the cathodes that were made of the equally mixed five types of cathode active materials to simulate the battery performance in the case that no separation happened in the battery recycling. This group is for the study of scenario III.

3.2. Half-Cell Fabrication

To study the performance of cathodes, several cathodes from each of the three groups were made. The cathode active material was acetylene black (MTI corporation), and the binder was PVDF (MTI corporation). The cathode active materials were mixed homogenously with PVDF and acetylene black with a weight ratio of 92:4:4, respectively. The solvent to make the cathode slurry was NMP (Sigma-Aldrich) that was mixed at a solid-to-liquid ratio of 1:2 before the cathode slurry was coated on the aluminum current collector. After making the cathodes, several coin half-cells (3032-type) were assembled in the order shown in Figure 3 in an argon-filled glovebox in the laboratory. Lithium foil was used as a reference electrode for all half-cells. The electrolyte used in the half-cells was 1 M LiPF$_6$ salt in 1:1 EC:DEC (by weight) solvent (Sigma-Aldrich).

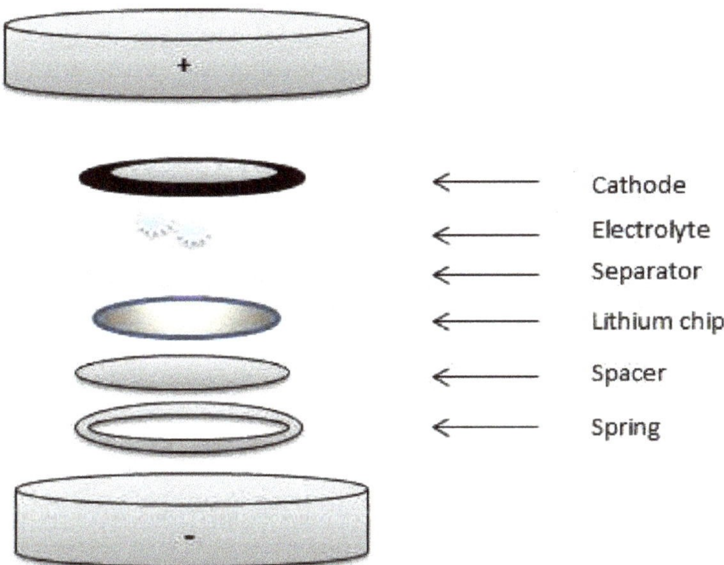

Figure 3. The schematic of the half-cell assembly of coin cells in glovebox.

4. Results and Discussion

In this section, the results of the computer simulation and experiments for scenarios I to III are presented. The results are generated based on one important assumption, that the efficiency of the regeneration process to recover the recycled cathode active materials is 100%. This means that the regeneration process is ideal and the capacity of the recycled active material after the regeneration reaches the capacity of the fresh material. This assumption was made to reduce the degree of freedom of the study to only focus on the effect of blending cathode materials on the cathode performance.

4.1. Scenario I

In this scenario, the separation process is assumed to be performed ideally. This means that a pure single regenerated cathode active material can be used to make new LIBs. The modeling was carried out for five of the most common cathode active materials: LCO, LMO, NCA, NMC, and LFP. The modeling parameters for each of the cathode active materials are listed in Table 3.

Table 3. Model parameters of the commonly used cathode active materials in LIBs [22–27].

Active Material	Density (kg/m^3)	D_{50} (μm)	Maximum State of Charge (mol/m^3)	Initial State of Charge (mol/m^3)	Reaction Rate COEFFICIENT, k_i (mol/s.m^2)	Theoretical Capacity (mAh/g)
LMO	4280	25	23,670.6	6000	5×10^{-10}	148.2
LCO	5050	12	51,555	23,750	1×10^{-7}	273.8
NMC	4770	10.5	51,385	21,500	1×10^{-11}	279.5
NCA	4450	13.6	46,319	8067.9	1×10^{-10}	278.9
LFP	3600	3.5	22,806	1000	3.63×10^{-11}	169.9

The open circuit voltage (OCV) profile of each cathode active material is shown in Figure 4. As seen, each material has a different nominal voltage. The OCV profile represents the ideal discharge performance of the half-cell.

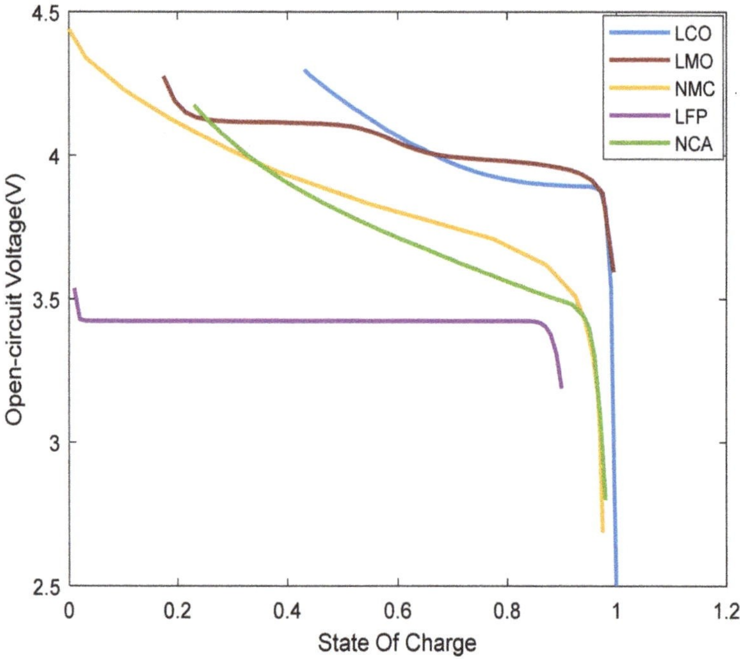

Figure 4. Open circuit voltage of some commercially available cathode active materials for LIBs versus the active material's state of charge at 25 °C.

The performance of half-cells with single cathode active materials at 1C discharge is shown in Figure 5. The actual capacity of the battery is lower than the theoretical capacity. The discrepancies of the actual and the theoretical capacity of cathode active materials are due to several factors. One of the factors is the discharge C-rate. As the C-rate increases, the capacity of the battery decreases [28]. In this study, we chose 1C because it is an appropriate indicator for the battery performance, especially for batteries used in hybrid electric vehicles and aircrafts. Another factor is the crystal structure of the active materials. As shown in Figure 5, the actual capacity of batteries, which is obtained from mathematical modeling, is in the range of the experimental results reported in the literature [29,30]. In addition, the modeling and experimental profiles are very close to each other, indicating the validity of the adopted mathematical model.

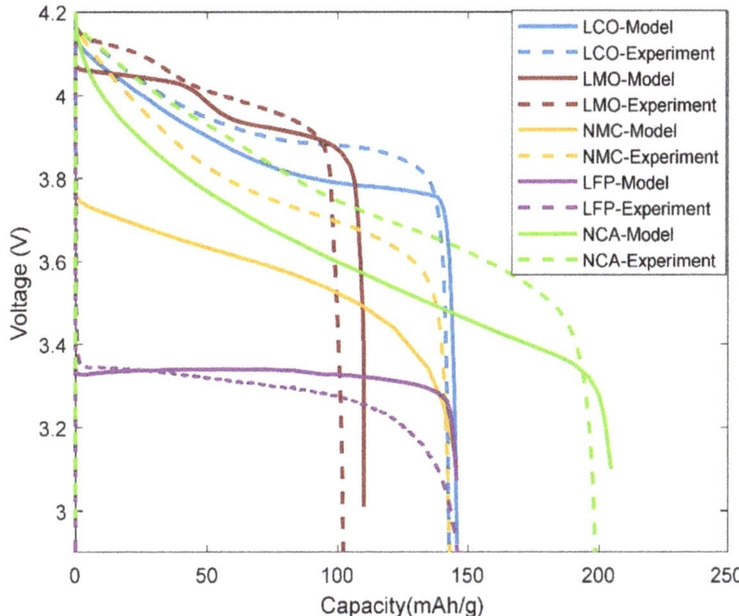

Figure 5. Discharge profile (1C and 25 °C) of half-cells made from cathodes with a single cathode active material.

4.2. Scenario II

In this scenario, the separation process is performed with an efficiency of less than 100% (actual separation process). Therefore, the single regenerated cathode active material contains a little of the other types of cathode active materials.

To obtain the electrochemical performance of the half-cells made from these regenerated cathode active materials, pure LCO, LMO, NMC, NCA, and LFP powders were mixed homogenously in different mass ratios. Five different mass ratios of 90%, 80%, 70%, 60%, and 50% were considered for the dominant active materials, and 2.5%, 5%, 7.5%, 10%, and 12.5% of each of the other four active materials were mixed with the dominant material, respectively. For example, for the dominant LCO cathode, we mixed 90 wt% LCO with 2.5 wt% NMC, 2.5 wt% LMO, 2.5 wt% NCA, and 2.5 wt% LFP for sample cathode one. For sample cathode two, we mixed 80 wt% LCO with 5 wt% NMC, 5 wt% LMO, 5 wt% NCA, and 5 wt% LFP. Similarly, we made sample cathodes three, four, and five for the dominant LCO of 70%, 60%, and 50%. We also carried out the same for the dominant LMO, NMC, NCA, and LFP cathode samples one to five. Obviously, in each half-cell, the dominant cathode active material has a higher percentage than the other four cathode active materials. After preparation of all samples and finishing the formation process, the half-cells were discharged between the voltage limits of 4.2 V and 2.9 V at a 1C rate and 25 °C.

Figure 6 shows the performance of the half-cells for five cathode active materials. As seen, when the NCA is the dominant active material, the capacity of the battery decreased with the decreasing percentage of NCA. The reduction in the capacity is because of the higher capacity of NCA among the other four types of the cathode active materials. When the LCO and NMC are the dominant materials, there is a slight reduction in the capacity when the percentage of the dominant active material reduces. This insignificant reduction of the capacity happens because these two cathode materials have a capacity in the average range among the other cathode active materials.

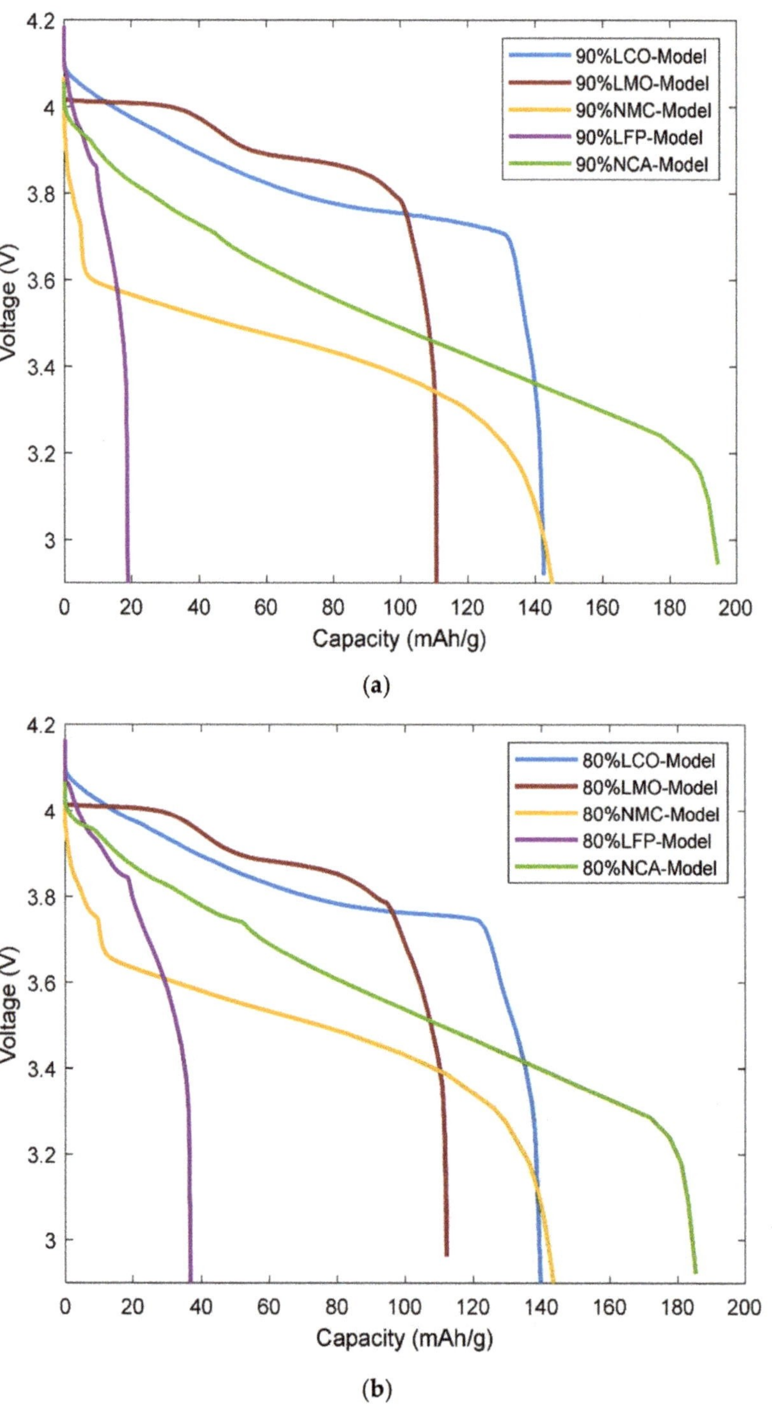

(**a**)

(**b**)

Figure 6. *Cont.*

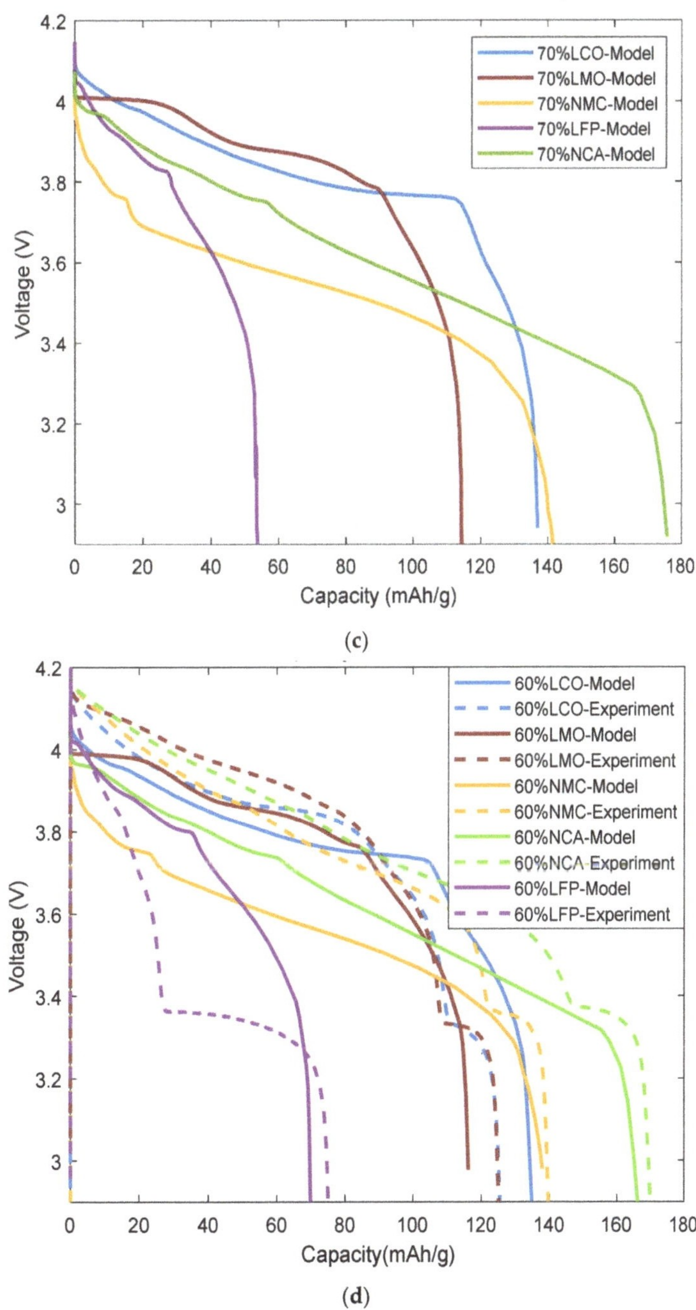

Figure 6. *Cont.*

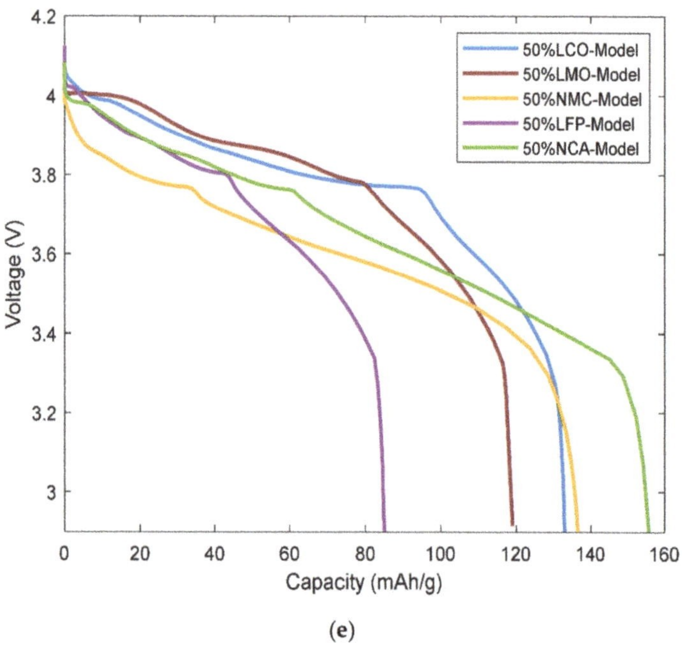

(e)

Figure 6. Discharge profile (1C and 25 °C) of half-cells made from a dominant cathode active material. (**a**) 90% dominant active material mixed with 10% other active materials (2.5% each), (**b**) 80% dominant active material mixed with 20% other active materials (5% each), (**c**) 70% dominant active material mixed with 30% other active materials (7.5% each), (**d**) 60% dominant active material mixed with 40% other active materials (10% each), and (**e**) 50% dominant active material mixed with 50% other active materials (12.5% each).

When the dominant material is LMO, the capacity of the half-cells is increased with the decreasing percentage of LMO. This increase is due to the difference between the capacity of the LMO and the other four cathode active materials. In fact, LMO has the lowest capacity among the other four materials, while it has the highest nominal voltage of 3.9 V.

The worst-case scenario was found when the dominant material is LFP. The capacity of the half-cells decreased significantly. This reduction of the capacity is due to the fact that the LFP has the lowest nominal voltage (3.2 V) compared to the other four cathode materials, as seen in Figure 4. Thus, the performance of the cathode with a dominant active material is dependent on the nominal voltage of the other mixed active materials. Therefore, all other active materials should be completely separated from the LFP.

For validation purposes, we made half-cells with 60% of the dominant cathode active material and 10% of each of the other four active materials and compared the modeling results with the experimental results. This comparison is shown in Figure 6d. As seen, there is a good agreement between the experiments and modeling in terms of the prediction of the cathode capacity. However, discrepancies are evident for the prediction of voltages, especially for the case when LFP is the dominant active material. Since our focus in this paper is the evaluation of the electrode capacity, the current model is valid. However, a more accurate model is required for studying the voltage trends.

4.3. Scenario III

In this scenario, the separation has not been performed during the recycling process, all types of cathode active materials are regenerated together, and a mixture of cathode active materials is used to make new LIBs. Three samples were prepared for this scenario.

For sample one, we made the half-cells from cathodes with equal weight percentages of the five cathode materials (20 wt% LCO, 20 wt% LMO, 20 wt% NMC, 20 wt% NCA, and 20 wt% LFP). Since from the results of scenario II we learned that LFP is not suitable to be in the mixture of cathode active materials, we made two more samples that have only 5 wt% and 0 wt% LFP. These two samples simulate the case that the separation is performed only for LFP. For sample two, we made the half-cells from cathodes with equal weight percentages of 23.75% for LCO, LMO, NMC, and NCA and 5 wt% LFP. For sample three, we made the half-cells from cathodes with equal weight percentages of 25% for LCO, LMO, NMC, and NCA. No LFP exists in sample three. The composition of samples one to three are listed in Table 4.

Table 4. Mass fraction of active materials in the cathode samples made for scenario III.

Cathode in the Half-Cells	LCO wt%	LMO wt%	NMC wt%	NCA wt%	LFP wt%
Sample 1 (20% each)	20	20	20	20	20
Sample 2 (5% LFP, 23.75% each)	23.75	23.75	23.75	23.75	5
Sample 3 (0% LFP, 25% each)	25	25	25	25	0

As seen in Figure 7, the best performance was found for the half-cells made from sample three, which has 0% LFP. The capacity of the half-cell made with a 5 wt% LFP (sample two) does not have noticeable a difference in comparison to the half-cell made from sample three. Therefore, we can conclude that a small fraction of LFP (typically less than 5 wt%) does not have a significant impact on the cathode capacity. However, the half-cell capacity decreases significantly when the weight percentage of LFP in the cathode active material mixture reaches 20% (sample one).

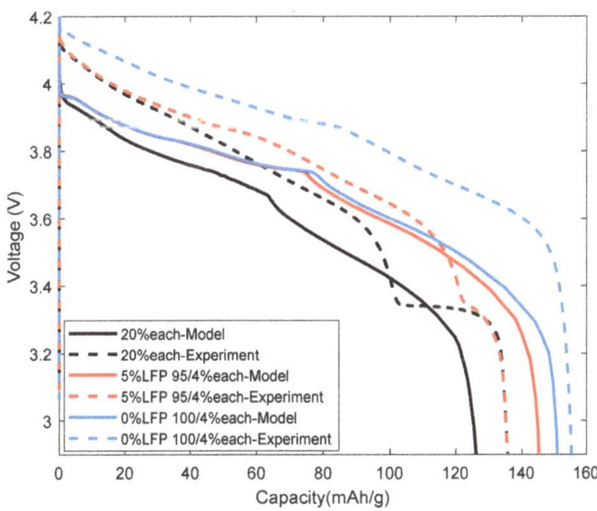

Figure 7. Discharge profile (1C, 25 °C) of cathodes made from the mixture of regenerated cathode active materials. Black curves: 20 wt% of each of the five types of cathode active materials (sample 1). Red curves: 5 wt% LFP and 23.75 wt% of each of the other four active materials (sample 2). Blue curves: 0 wt% LFP and 25 wt% of each of the other four cathode active materials (sample 3).

5. Conclusions

The performance of cathodes made from recycled cathode active materials for three scenarios of the active materials' separation was studied. In scenario one, the separation process in the recycling facility was ideal, and a pure single cathode active material was used to make new batteries. In scenario two, the separation process for active materials in the recycling facility was realistic, with an efficiency of less than 100%. In this scenario, a single cathode active material that contains a little of the other types of cathode active materials was used to make new batteries. In scenario three, the separation was not performed, and all types of cathode active materials were regenerated together and used to make new batteries. The results were generated through modeling and computer simulation and several experiments were conducted for validation of the model. The cathode active materials that were studied are five commercially available cathode active materials: LMO, LCO, NMC, NCA, and LFP. The results indicated that the fabrication of new LIBs with a mixture of regenerated cathode active materials is possible when the cathode active materials are not separated from each other. However, it is better that the separation process is added to the recycling process, especially for minimization of the amount of LFP in the mixture of cathode active materials. It is not necessary that the separation process is ideal. However, more studies are still required to show that making new LIBs with a dominant cathode active material is technically and economically feasible.

Author Contributions: Conceptualization, S.F.; methodology, S.F.; software, H.A.-S.; validation, H.A.-S.; formal analysis, H.A.-S., and S.F.; investigation, H.A.-S., and S.F.; resources, S.F.; data curation, H.A.-S.; writing—original draft preparation, H.A.-S.; writing—review and editing, S.F.; visualization, H.A.-S.; supervision, S.F.; project administration, S.F.; funding acquisition, S.F. All authors have read and agreed to the published version of the manuscript.

Funding: This research received no external funding.

Acknowledgments: The authors gratefully acknowledge supports from the University of Akron and Al Jouf University.

Conflicts of Interest: The authors declare no conflict of interest.

Nomenclature

A	Apparent surface area of electrode (m^2)
c	Concentration (mol/m^3)
D	Diffusivity (m^2/s)
D_{50}	Mean diameter of particles (m)
E_{eq}	Equilibrium potential (V)
f	Mean molar activity coefficient of inorganic salt in electrolyte
F	Faraday's constant (96,485 C/mol)
i	Currant density (A/m^2)
$I_{applied}$	Applied current to battery (A)
i_0	Exchange currant density (A/m^2)
i_{local}	Local current density generation (A/m^2)
k	Electrochemical reaction rate coefficient (m/s)
L	Thickness (m)
N	Flux ($mol/m^2 \cdot s$)
Q	Capacity (Ah/m^2)
r	Radial direction in spherical coordinate system (m)
R	Universal gas constant (8.314 J/mol K)
r_p	Average radius of active material particles
S_a	Surface area per unit volume (m^2/m^3)
SOC	State of charge of the active material
t	time (s)
t^0_+	Lithium-ion transference number
T	Temperature (K)
x	x direction in cartesian coordinate system (m)
Y	Mole fraction of an active material in the mixture of active materials

Greek Letters

ε	Volume fraction
η	Polarization (V)
σ	Conductivity (S/m)
ϕ	Potential (V)

Subscripts

0	Initial
i	Index for ith cathode active materials in the cathode mixture
l	Liquid phase (electrolyte)
cat	Cathode
eff	Effective
max	maximum
min	minimum
pos	Positive electrode (cathode)
ref	Reference
s	Solid phase of electrode (active/conductive material)
sep	Separator
surf	Surface

Abbreviations

DEC	Dimethyl carbonate
EC	Ethylene carbonate
LCO	$LiCoO_2$
LFP	$LiFePO_4$
LIB	Lithium-ion battery
LMO	$LiMn_2O_4$
NCA	$LiNi_xCo_yAl_{(1-x-y)}O_2$
NMC	$LiNi_xCo_yMn_{(1-x-y)}O_2$
NMP	N-Methyl-2-pyrrolidone
OCV	Open circuit voltage
PVDF	polyvinylidene fluoride

References

1. Zubi, G.; Dufo-López, R.; Carvalho, M.; Pasaoglu, G. The lithium-ion battery: State of the art and future perspectives. *Renew. Sustain. Energy Rev.* **2018**, *89*, 292–308. [CrossRef]
2. Lu, L.; Han, X.; Li, J.; Hua, J.; Ouyang, M. A review on the key issues for lithium-ion battery management in electric vehicles. *J. Power Sources* **2013**, *226*, 272–288. [CrossRef]
3. Olivetti, E.A.; Ceder, G.; Gaustad, G.G.; Fu, X. Lithium-Ion Battery Supply Chain Considerations: Analysis of Potential Bottlenecks in Critical Metals. *Joule* **2017**, *1*, 229–243. [CrossRef]
4. An, S.J.; Li, J.; Daniel, C.; Mohanty, D.; Nagpure, S.; Wood, D.L. The state of understanding of the lithium-ion-battery graphite solid electrolyte interphase (SEI) and its relationship to formation cycling. *Carbon* **2016**, *105*, 52–76. [CrossRef]
5. Zhang, X.; Lu, C.; Peng, H.; Wang, X.; Zhang, Y.; Wang, Z.; Zhong, Y.; Wang, G. Influence of sintering temperature and graphene additives on the electrochemical performance of porous Li4Ti5O12 anode for lithium ion capacitor. *Electrochim. Acta* **2017**, *246*, 1237–1247. [CrossRef]
6. Wu, B.; Wang, J.; Li, J.; Lin, W.; Hu, H.; Wang, F.; Zhao, S.; Gan, C.; Zhao, J. Morphology controllable synthesis and electrochemical performance of $LiCoO_2$ for lithium-ion batteries. *Electrochim. Acta* **2016**, *209*, 315–322. [CrossRef]
7. Fergus, J.W. Recent developments in cathode materials for lithium ion batteries. *J. Power Sources* **2010**, *195*, 939–954. [CrossRef]
8. Plichta, E.; Behl, W. A low-temperature electrolyte for lithium and lithium-ion batteries. *J. Power Sources* **2000**, *88*, 192–196. [CrossRef]
9. Chen, J.; Wang, S.; Ding, L.; Jiang, Y.; Wang, H. Performance of through-hole anodic aluminum oxide membrane as a separator for lithium-ion battery. *J. Membr. Sci.* **2014**, *461*, 22–27. [CrossRef]
10. Arora, P.; Zhang, Z. Battery Separators. *Chem. Rev.* **2004**, *104*, 4419–4462. [CrossRef]
11. Ritchie, A.; Howard, W. Recent developments and likely advances in lithium-ion batteries. *J. Power Sources* **2006**, *162*, 809–812. [CrossRef]
12. de la Llave, E.; Talaie, E.; Levi, E.; Nayak, P.; Dixit, M.; Rao, P.T.; Hartmann, P.; Chesneau, F.; Major, D.; Greenstein, M.; et al. Improving Energy Density and Structural Stability of Manganese Oxide Cathodes for Na-Ion Batteries by Structural Lithium Substitution. *Chem. Mater.* **2016**, *28*, 9064–9076. [CrossRef]
13. Albertus, P.; Christensen, J.; Newman, J. Experiments on and Modeling of Positive Electrodes with Multiple Active Materials for Lithium-Ion Batteries. *J. Electrochem. Soc.* **2009**, *156*, A606. [CrossRef]

14. Jeong, S.K.; Shin, J.S.; Nahm, K.S.; Kumar, P.; Stephan, A.M. Electrochemical studies on cathode blends of $LiMn_2O_4$ and $Li[Li1/15Ni1/5Co_2/5Mn1/3O_2]$. *Mater. Chem. Phys.* **2008**, *111*, 213–217. [CrossRef]
15. Jobst, N.M.; Hoffmann, A.; Klein, A.; Zink, S.; Wohlfahrt-Mehrens, M. Ternary Cathode Blend Electrodes for Environmentally Friendly Lithium-Ion Batteries. *ChemSusChem* **2020**, *13*, 3928–3936. [CrossRef]
16. Al-Shammari, H.; Esmaeeli, R.; Aliniagerdroudbari, H.; Alhadri, M.; Hashemi, S.R.; Zarrin, H.; Farhad, S. Recycling lithium-ion battery: Mechanical separation of mixed cathode active materials. In *ASME International Mechanical Engineering Congress and Exposition*; Proceedings (IMECE); American Society of Mechanical Engineers (ASME): Salt Lake City, UT, USA, 2019. [CrossRef]
17. Al-Shammari, H.; Farhad, S. Regeneration of Cathode Mixture Active Materials Obtained from Recycled Lithium Ion Batteries. SAE Technical Paper 2020-01-0864. 2020. Available online: https://www.sae.org/publications/technical-papers/content/2020-01-0864/ (accessed on 3 December 2021).
18. Al-Shammari, H.; Farhad, S. Heavy liquids for rapid separation of cathode and anode active materials from recycled lithium-ion batteries. *Resour. Conserv. Recycl.* **2021**, *174*, 105749. [CrossRef]
19. Kay, I.; Esmaeeli, R.; Hashemi, S.R.; Mahajan, A.; Farhad, S. Recycling Li-ion batteries: Robot disassembly of electric vehicle battery systems. In Proceedings of the ASME 2019, Internal Mechanical Engineering Congress and Exposition (IMECE), Salt Lake City, UT, USA, 8–14 November 2019. [CrossRef]
20. Doyle, M.; Newman, J. The use of mathematical modeling in the design of lithium/polymer battery systems. *Electrochim. Acta* **1995**, *40*, 2191–2196. [CrossRef]
21. Chabot, V.; Farhad, S.; Chen, Z.; Fung, A.S.; Yu, A.; Hamdullahpur, F. Effect of electrode physical and chemical properties on lithium-ion battery performance. *Int. J. Energy Res.* **2013**, *37*, 1723–1736. [CrossRef]
22. Liu, C.; Li, H.; Kong, X.; Zhao, J. Modeling analysis of the effect of battery design on internal short circuit hazard in $LiNi0.8Co0.1Mn0.1O2/SiOx$-graphite lithium ion batteries. *Int. J. Heat Mass Transf.* **2020**, *153*, 119590. [CrossRef]
23. Miranda, D.; Gören, A.; Costa, C.; Silva, M.; Almeida, A.; Lanceros-Mendez, S. Theoretical simulation of the optimal relation between active material, binder and conductive additive for lithium-ion battery cathodes. *Energy* **2019**, *172*, 68–78. [CrossRef]
24. Ramadass, P.; Haran, B.; White, R.; Popov, B.N. Mathematical modeling of the capacity fade of Li-ion cells. *J. Power Sources* **2003**, *123*, 230–240. [CrossRef]
25. Farhad, S.; Nazari, A. Introducing the energy efficiency map of lithium-ion batteries. *Int. J. Energy Res.* **2019**, *43*, 931–944. [CrossRef]
26. Paul, N.; Keil, J.; Kindermann, F.M.; Schebesta, S.; Dolotko, O.; Mühlbauer, M.J.; Kraft, L.; Erhard, S.; Jossen, A.; Gilles, R. Aging in 18650-type Li-ion cells examined with neutron diffraction, electrochemical analysis and physico-chemical modeling. *J. Energy Storage* **2018**, *17*, 383–394. [CrossRef]
27. Dong, T.; Peng, P.; Jiang, F. Numerical modeling and analysis of the thermal behavior of NCM lithium-ion batteries subjected to very high C-rate discharge/charge operations. *Int. J. Heat Mass Transf.* **2018**, *117*, 261–272. [CrossRef]
28. Tang, W.; Liu, L.; Tian, S.; Li, L.; Yue, Y.; Wu, Y.; Guan, S.; Zhu, K. Nano-$LiCoO_2$ as cathode material of large capacity and high rate capability for aqueous rechargeable lithium batteries. *Electrochem. Commun.* **2010**, *12*, 1524–1526. [CrossRef]
29. Qiu, L.; Shao, Z.; Wang, D.; Wang, W.; Wang, F.; Wang, J. Enhanced electrochemical properties of LiFePO4 (LFP) cathode using the carboxymethyl cellulose lithium (CMC-Li) as novel binder in lithium-ion battery. *Carbohydr. Polym.* **2014**, *111*, 588–591. [CrossRef]
30. Nishi, Y. Lithium ion secondary batteries; past 10 years and the future. *J. Power Sources* **2001**, *100*, 101–106. [CrossRef]

 *energies*

Article

Recycling and Reusing Copper and Aluminum Current-Collectors from Spent Lithium-Ion Batteries

Hamid Khatibi [1,2], Eman Hassan [1], Dominic Frisone [1], Mahdi Amiriyan [2], Rashid Farahati [2] and Siamak Farhad [1,*]

1   Advanced Energy and Manufacturing Laboratory, Department of Mechanical Engineering, University of Akron, Akron, OH 44325, USA
2   Schaeffler Group, Wooster, OH 44691, USA
*   Correspondence: sfarhad@uakron.edu

Abstract: The global transition to electric vehicles and renewable energy systems continues to gain support from governments and investors. As a result, the demand for electric energy storage systems such as lithium-ion batteries (LIBs) has substantially increased. This is a significant motivator for reassessing end-of-life strategies for these batteries. Most importantly, a strong focus on transitioning from landfilling to an efficient recycling system is necessary to ensure the reduction of total global emissions, especially those from LIBs. Furthermore, LIBs contain many resources which can be reused after recycling; however, the compositional and component complexity of LIBs poses many challenges. This study focuses on the recycling and reusing of copper (Cu) and aluminum (Al) foils, which are the anode and cathode current-collectors (CCs) of LIBs. For this purpose, methods for the purification of recycled Cu and Al CCs for reusing in LIBs are explored in this paper. To show the effectiveness of the purification, the recycled CCs are used to make new LIBs, followed by an investigation of the performance of the made LIBs. Overall, it seems that the LIBs' CCs can be reused to make new LIBs. However, an improvement in the purification method is still recommended for future work to increase the new LIB cycling.

Keywords: lithium-ion battery; current-collectors; recycling and reusing; copper; aluminum

Citation: Khatibi, H.; Hassan, E.; Frisone, D.; Amiriyan, M.; Farahati, R.; Farhad, S. Recycling and Reusing Copper and Aluminum Current-Collectors from Spent Lithium-Ion Batteries. *Energies* 2022, 15, 9069. https://doi.org/10.3390/en15239069

Academic Editor: Seung-Wan Song

Received: 2 November 2022
Accepted: 27 November 2022
Published: 30 November 2022

Publisher's Note: MDPI stays neutral with regard to jurisdictional claims in published maps and institutional affiliations.

1. Introduction

The electric vehicle (EV) revolution, driven by the imperatives to decarbonize personal and commercial transportation to meet global targets of greenhouse gas reduction and air quality improvements in urban centers, is set to change the automotive industry radically. In 2022, sales of electrified vehicles reached a 10% market share with 6 million vehicles sold. This is double that of 2021 and quadrupling 2020 sales, evidencing ongoing support for the transition [1].

With the ever-growing need for lithium-ion battery (LIBs) technology, many LIBs are destined for retirement in the coming years [2–4]. Historically, LIBs especially for EV applications have suffered incredibly low recycling rates [5]. The complex, low-profit material processing procedures and lack of consumer value propositions for recycling old electronics resulted in low interest in recycling LIB cells and such batteries ended up in landfills. The lack of proper disposal of spent LIBs results in grave ecological consequences [6]. While this problem still seems to persist for consumer electronics [7,8], the residual value, economics of scale, material shortage and need for disposal associated with EV battery packs provide an economic incentive to recycle LIB cells. Thus, recycling of spent LIBs has received substantial attention in recent years [9–14]. However, some researchers suggest that the retired LIBs from applications such as EVs and electric aircraft can be used in applications such as renewable storage before recycling [15,16]. Although new materials are being developed for next-generation LIBs [17] to overcome the safety issue of LIBs [18] and reduce the use of precious materials in electrochemical systems [19], recycling will be still a need to return the battery materials back to the supply chain and the LIB manufacturing line. The forecasts show that in the coming decades, tens of millions of EVs will be produced annually. Careful farming of the resources used in EV battery manufacturing will be essential to ensure the sustainability of the automotive industry in the future, ensuring material and energy-efficient 3R systems (reduce, reuse, recycle).

In the waste management hierarchy, reusing materials is considered preferable over recycling them, to extract maximum economic value and minimize environmental impacts. Many companies in various parts of the world are already piloting the second use of EV LIB cells for a range of storage applications [20–22]. Today, advanced sensors and improved methods of monitoring battery packs and cells in the field and end-of-life testing are used. This would enable the characteristics of individual end-of-life batteries to be better matched to proposed second-use applications, with affiliated advantages in the lifetime, safety, and market value [23,24]. The influences of retired LIB packs and subsequently cells from EVs on resource conservation and environmental protection will be positive. Any national or regional end-of-life (EOL) strategy will need to account for the reduced demand for energy storage as well as the reduced supply of EOL batteries [25]. Questions regarding the economic viability and safety of second-life batteries are being quickly answered both practically and theoretically. They will undoubtedly play a role in the clean energy transition.

Although EVs have no point emission during operation, many factors related to LIB manufacturing, use, and recycling heavily influence the true environmental impact [26–28]. With the market expansion of EVs in recent and coming years, LIB manufacturing and upstream industries have increased substantially. Thus, consequently creating environmental burdens, such as resource consumption, energy generation, and wastes emission (including gaseous, liquid, and solid wastes) [29,30]. The high efficiency of electric powertrains decreases the impact of the electrical source, but global clean energy generation is still required to meaningfully impact global emissions [31]. Recycling, which can be modeled as either a downstream or upstream step of the manufacturing process, has non-negligible environmental impacts. Meaningful research works have been published, modeling many recycling processes' general efficiency and emissions [32–34]. Yet, the rate of change within the industry necessitates a constant reexamination of assumptions and calculations.

The ever-changing LIB cell composition is one of the many categorical challenges to creating robust, flexible recycling systems. The intense competition for high-performance LIB cells has inspired substantial research across all battery components. A historically under-researched component, which is largely unaccounted for in the analysis of recycling methods, is the current-collector (CC). CCs serve a vital bridge function in supporting the active materials, binders, and conductive additives, as well as electrically connecting the anodes and cathodes to the external circuit. High-purity copper (Cu) and aluminum (Al) foils are predominately used as CCs for anodes and cathodes, respectively. Recently, various factors of CCs such as the thickness, hardness, compositions, coating layers, and structures have been modified to improve aspects of battery performance such as the charge/discharge cyclability, energy density, and the rate performance of a cell [35,36]. Lithium ions and electrons should move rapidly to and from the anode and cathode particle surfaces to charge and discharge the cell with a high current density. In addition, CCs should have high mechanical strength, chemical and electrochemical stability, and adhesion between the active material layer and the CC surface. To realize these requirements, the optimization of materials for CCs, the structural modification of CCs, and the formation of a surface layer on CCs have been performed [37–40].

The established recycling processes are focused mainly on the recovery of cathode active materials [41–47]. Even though it is a small weight percentage of the cell, it comprises a large portion of the material's elemental value. In addition to the cathode materials, research in recycling CCs can incentivize the recycling process, especially, if the recycling process becomes automated [48,49].

Tight and closed loops for all battery components will be necessary to support production at the global electric vehicle market scale. Towards that effort, this research work examines the efficacy of reusing Al and Cu CCs in three steps: (a) impurity analysis of scrap material after ultrasonic solvent bath, (b) impurity analysis after thermal melting process, and finally (c) impurity analysis after reusing the recycled CCs in half-cells.

2. Materials and Methods

2.1. Separation of Al and Cu CCs

Figure 1a shows the fluff materials that were received from a LIB recycling facility. The fluff materials consist of the Al CC, Cu CC, plastics separator, battery casing, and battery Table Figure 1b,c show the Al and Cu CCs, respectively, after separation from the fluff materials. After the separation, the cleaning/purification step starts.

(**a**)   (**b**)   (**c**)

Figure 1. Material obtained from shredded scrap, (**a**) fluff, (**b**) separated Al, (**c**) separated Cu.

2.2. Ultrasonic Cleaning and Characterization of Surface and Bulk Impurities

The effectiveness of N-methyl-2-pyrrolidone (NMP), distilled water (RO), and ethanol as solutions for cleaning surface impurities in Al and Cu CCs obtained from recycled batteries was tested. These solutions were selected because they were accessible in the lab. Researchers may try other solutions which are more effective than these solutions. Al and Cu obtained from shredded battery cells were cleaned by placing them in an ultrasonic cleaner for 4 min at room temperature. Washed samples using the three cleaning methods were analyzed using scanning electron microscopy (SEM) (FEI Perception V4.6) and the energy-dispersive x-ray spectroscopy (EDX) method. To detect impurities on and below the surface, the SEM acceleration voltage was adjusted to 10 keV and 20 keV, respectively. The 20 keV X-ray beam energy seems to be high enough to penetrate a couple of tens of microns into the sample. Unwashed shredded battery cells containing Al and Cu were analyzed in the same way for control.

2.3. Melting, Molding, and Characterization of Surface and Bulk Impurities

The second phase of this study involved recycling the washed Al and Cu CCs into new battery cells and components. As before, Al and Cu are separated from the used LIBs and washed. To-be-recycled materials were placed in graphite crucibles and melted using a 1500-watt tabletop furnace (Tabletop Furnace Company, Tacoma, WA) with a maximum heating temperature of 1205 °C (2200 °F) and a standard 15 Amp circuit as seen in Figure 2a. Molten Al and Cu are then poured into graphite molds to form several 1 mm thick disks of 10 mm diameter Figure 2b. After Al and Cu solidification, the disks were pressed to be 20 mm in diameter using a DAKE-50 Tone press (Figure 2c) and then polished as seen in Figure 2d.

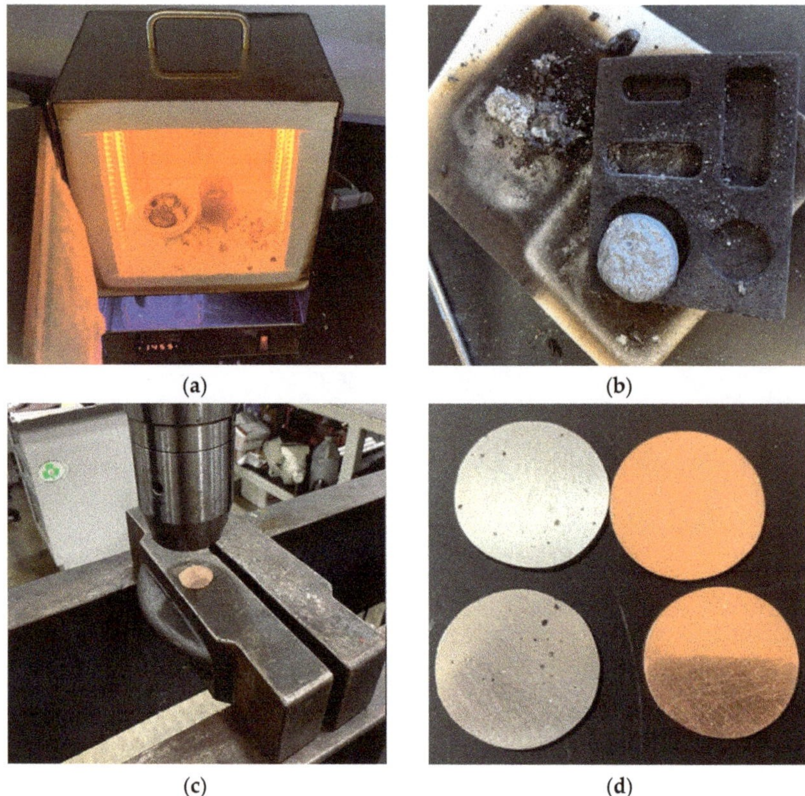

Figure 2. (**a**) Tabletop furnace used for melting, (**b**) mold for forming disks from molten material, (**c**) press for forming the samples (DAKE-50 Tone), and (**d**) polished and pressed CC disks.

During melting, the metal which enters the furnace is not all used in the final product and some of it is lost. Consequently, melting efficiency is a key parameter that must be accounted for in the results. It can be quantified as the weight ratio of output material to input material. Furthermore, the dimensions for preparing the CCs were chosen to improve the melting efficiency. Additionally, compression was utilized to prevent oxidation as it reduces the CC surfaces which are in contact with oxygen. It is noted that the melting efficiency in our study is low (less than 50%) due to the limitation of the equipment and technology of melting that we have in the lab. Large-scale recycling facilities can be equipped with equipment and technologies so that the melting efficiency increases to close to 100%.

As seen in Figure 1b, Al material acquired from crushing is in a bullet-like form with an apparently high surface melting point. This has made cleaning cathode active material from the surface difficult and would therefore lead to impurities at the grain boundaries. The issue of clumping did not occur in the crushed Cu material which is favorable for cleaning and improving melting efficiency.

The contamination effect of graphite crucibles on the melting process should be noted. Therefore, samples were analyzed through SEM and EDX to observe the implications of melting on the purity of the CCs. SEM images were taken linearly to show up to 400 microns on the surface of each sample. The acceleration voltage was kept at 10 keV. The percentage of impurities on the surface was measured using EDX for three spots and the average value was converted to weight percentage. EDX was the accessible characterization method for this study. Other methods such as ICP-MS can be sought out in future studies to assess elemental composition more accurately.

2.4. Half-Cell Construction and Final Characterization of Surface Impurities

Using the Al and Cu CCs from the second phase of this study, cathodes and anodes were fabricated and tested in four half-cells. Cathodes were made by mixing $LiNi_{1/3}Mn_{1/3}Co_{1/3}O_2$ (NMC111) with polyvinylidene fluoride (PVDF) binder and acetylene black as the conductive material in a ratio of 90/5/5 weight percent, respectively, as shown in Figure 3a. The NMC was mixed with NMP in a Thinky centrifugal mixer (Thinky U.S.A., Inc., Laguna Hills, CA) at 1000 rpm for 2 min. The binder and conductive material were added gradually using the same mixing program. Using a similar procedure, anodes were made by mixing graphite with carboxymethyl cellulose (CMC) and styrene-butadiene rubber (SBR) binders and acetylene black as the conductive material in a ratio of 89/7/4 weight percent, respectively, as shown in Figure 3b.

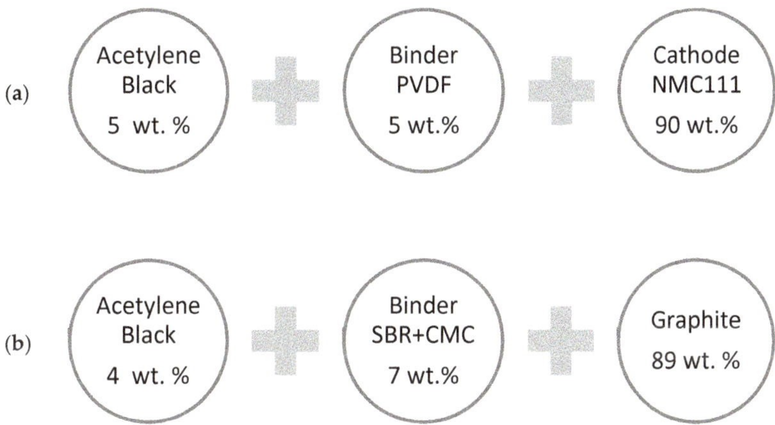

Figure 3. Schematic representation of materials mixed in (**a**) cathode and (**b**) anode.

Four coin-cell type half-cells were assembled using the anodes and cathodes made from recycled materials along with half-cells made from new electrode material. Half-cell components and coin-cell assembly can be seen in Figure 4. Coin-cells were assembled by placing punched cathodes or anodes, polypropylene (PP)/polyethylene (PE)/polypropylene (PP) tri-layer separator, then Li metal chip in coin-cell casings along with a wave spring. Each half-cell was tested at two C-rates of C/5 and C/10.

Aluminum Current-Collector	Copper Current-Collector
Cathode (Ni-Mn-Co)	Anode (Graphite)
Separator PP/PE/PP	Separator PP/PE/PP
Li Metal	Li Metal
Stainless Steel Spacer	Stainless Steel Spacer
Wave Spring	Wave Spring
(a)	(b)

Figure 4. The assembly of half-cell LIBs, (**a**) cathode and (**b**) anode.

To analyze the effect of chemical reactions during battery cycling on the CCs' chemical compositions, the half-cells were opened after cycling to observe the reappearance of surface impurities on the CCs' after testing using SEM and EDX.

3. Results and Discussion

3.1. Surface Impurities of Cleaned CCs by the Ultrasonic Technique

For the first phase of this study, the impurity reduction of washed CCs was analyzed. The EDX results by elemental weight percentages of the anode can be found in Figure 5. Two measurements on and below the surface were taken from four anode CC samples after NMP, RO, and ethanol cleaning, and one as-received sample as a control. A summary of the result can be seen in Figure 6.

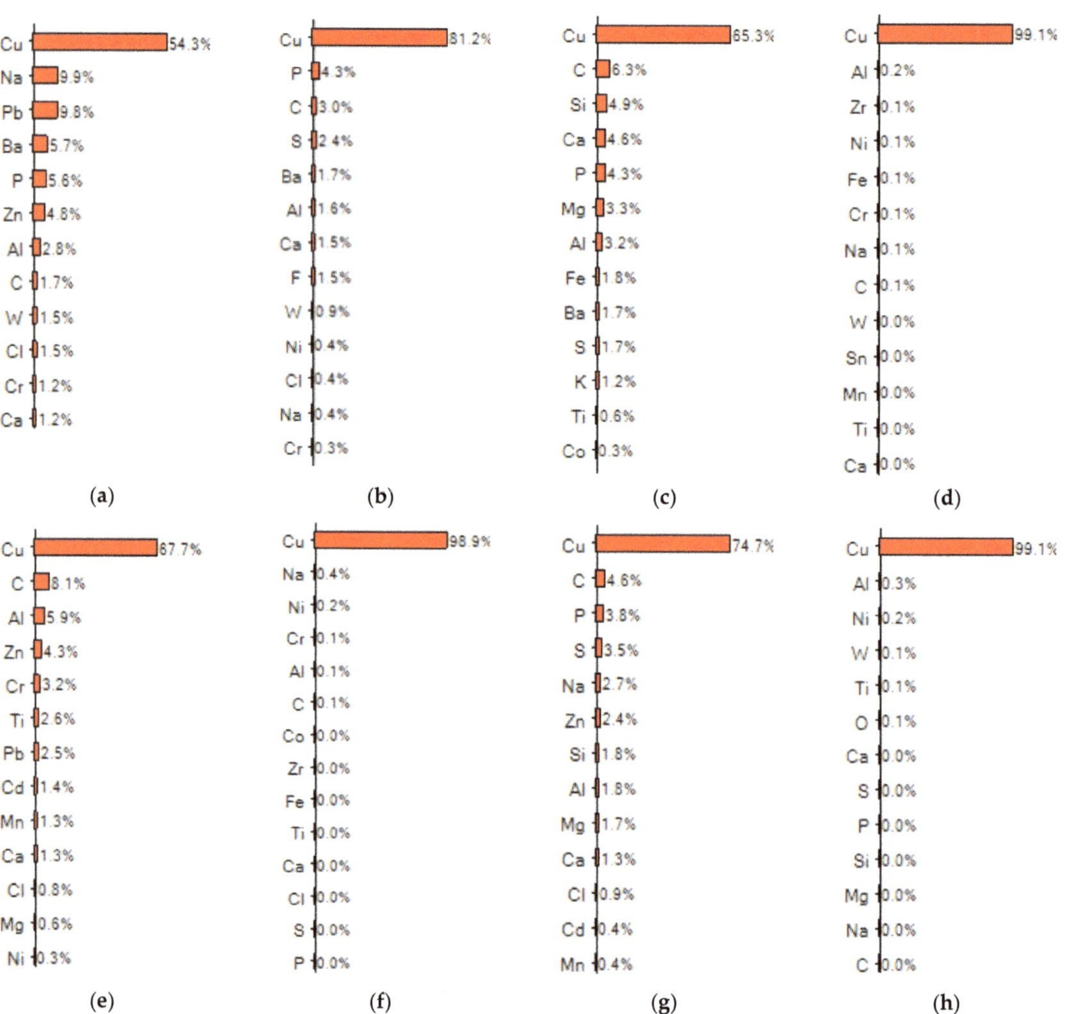

Figure 5. The EDX results for recycled Cu CCs show the composition (**a**) on the surface as-received without cleaning, (**b**) below the surface as-received without cleaning, (**c**) on the surface after cleaning by NMP, (**d**) below the surface after cleaning by NMP, (**e**) on the surface after cleaning by RO, (**f**) below the surface after cleaning by RO, (**g**) on the surface after cleaning by ethanol, and (**h**) below the surface after cleaning by ethanol.

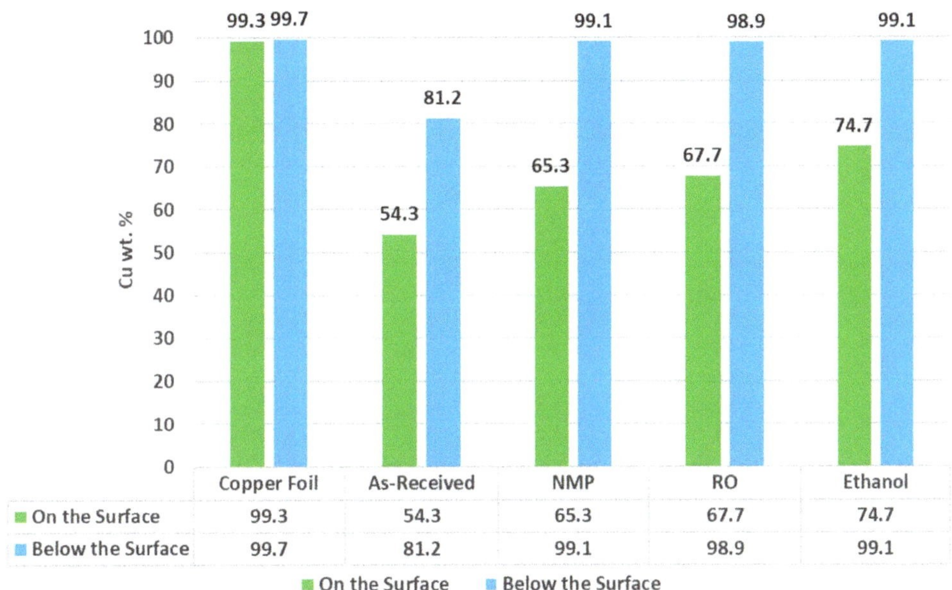

Figure 6. The weight percentage of Cu on and below the surface of a LIB-graded fresh Cu foil, recycled Cu as-received from spent batteries, and recycled Cu after washing with NMP, RO, and ethanol.

The as-received Cu foils showed 45.7 wt.% of impurities on the surface. For reference, new Cu foil supplied to manufacturing facilities contains only 0.7 wt.% of surface impurities. The most effective solution for washing the electrode surface was ethanol (77.4%), followed by RO (67.7%), with NMP being the least effective (65.3%). As expected, impurities below the surface were much less than those above the surface with 18.8 wt.% for as-received and ~1 wt.% for washed samples. Ethanol and NMP were most effective at washing impurities below the surface. In contrast to impurities on the surface, CCs washed with NMP showed fewer impurities than those washed with RO.

The data strongly indicates many impurities are present on and below the surface of used CCs. The impurities are due to the aggressive electrochemical environment inside a battery with potential contribution coming from shredding after retirement. These impurities must be washed before proceeding to the melting phase of the recycling process. As shown previously, the best solution for cleaning the surface of recycled Cu is ethanol, which removes up to 25.3% of surface impurities. Similar results were found for reducing impurities in Al CCs. Thus, all CCs subjected to melting were first cleaned with ethanol for 4 min at 24 °C (75 °F) in an ultrasonic cleaner before melting.

3.2. Surface Impurities of CCs after Melting and Battery Testing

Results from the SEM and elemental assessment of impurities on the surface of Al collectors are shown in Figures 7 and 8. SEM images of Al CCs at three relevant stages for assessment: as-received, after melting, and after battery testing, are shown in Figure 8a–c show. For these Al CCs, three impurities were found comprising Cu, manganese, and nickel. Overall, Al purity increased by 2.2 wt.% after recycling and decreased only by 1.1 wt.% after battery testing.

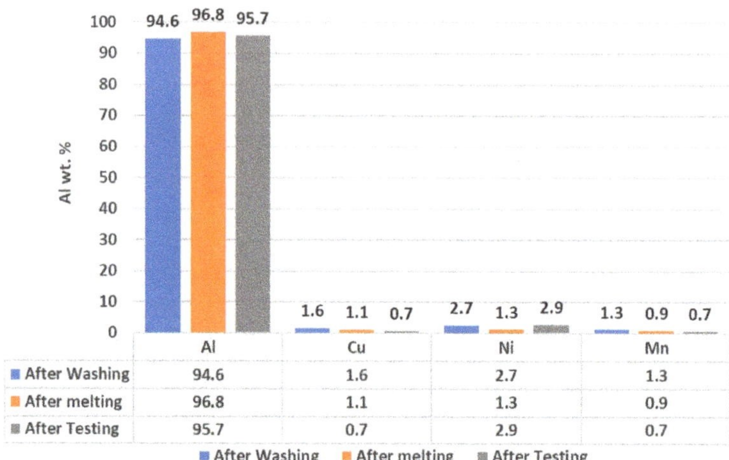

Figure 7. EDX test results on the surface of Al CCs after washing, melting, and battery testing.

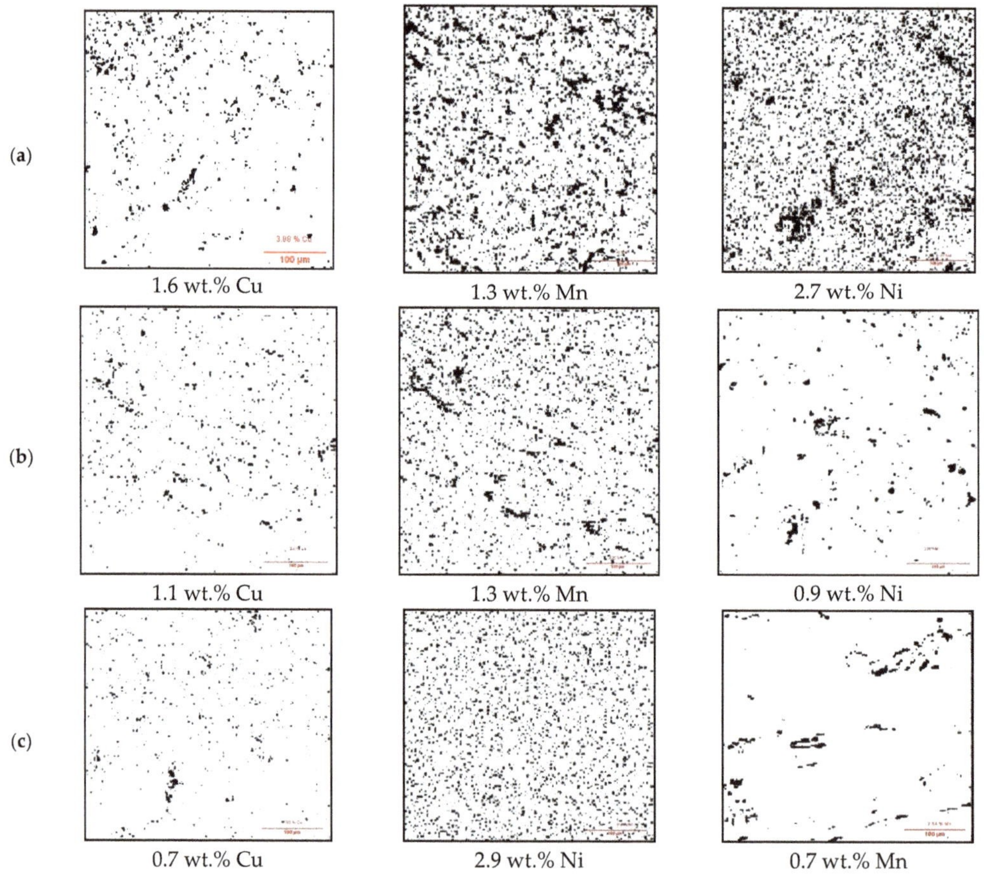

	Al	Cu	Ni	Mn
After Washing	94.6	1.6	2.7	1.3
After melting	96.8	1.1	1.3	0.9
After Testing	95.7	0.7	2.9	0.7

(**a**) 1.6 wt.% Cu   1.3 wt.% Mn   2.7 wt.% Ni

(**b**) 1.1 wt.% Cu   1.3 wt.% Mn   0.9 wt.% Ni

(**c**) 0.7 wt.% Cu   2.9 wt.% Ni   0.7 wt.% Mn

Figure 8. Elemental mapping for (**a**) As-received Al CC from used batteries, (**b**) Al CC from recycled batteries after melting, and (**c**) Al CC from recycled batteries after reuse in new battery and cycling.

It is important to note that all impurities decreased after the thermal treatment of the as-received collectors. Cu content decreased by 0.5 wt.%, nickel by 1.4 wt.%, and manganese by 0.4 wt.%. This drop in impurities proves that the recycling process is capable of reversing impurities in the CCs. Cu and manganese impurities continued to decrease by 0.4 and 0.2 wt.%, respectively, after battery operation. However, nickel content increased by 1.6 wt.%. The increase in nickel impurity after battery operation indicates that some CC degradation occurs along with the loss of active cathode material. Nickel content in impurities is higher than all other elements, especially for as-received CCs (2.7 wt.%) and recycled CCs used in battery testing (2.9 wt.%).

The SEM and elemental assessment results for Cu CCs at the three relevant stages are shown in Figures 9 and 10a–c. The impurities found on Cu CCs were Al and nickel. The purity of Cu showed only a slight increase of 0.3 wt.% after recycling and a very small decrease of 0.1 wt.% after battery testing. After recycling, Al impurities decreased by 0.4 wt.%. Contrastingly, nickel impurities increased slightly by 0.1 wt.%. Battery testing showed no impact on Al content quantities and only slightly increased nickel content by another 0.1 wt.%.

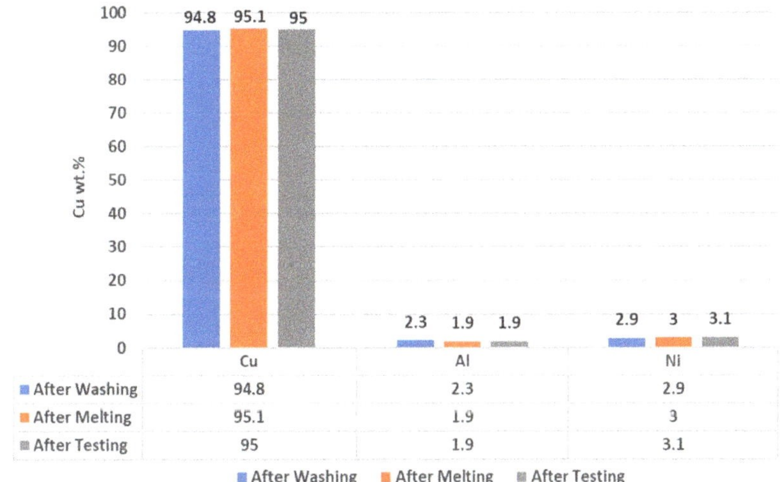

	Cu	Al	Ni
After Washing	94.8	2.3	2.9
After Melting	95.1	1.9	3
After Testing	95	1.9	3.1

Figure 9. EDX test results on the surface of Cu CC after washing, melting, and battery testing.

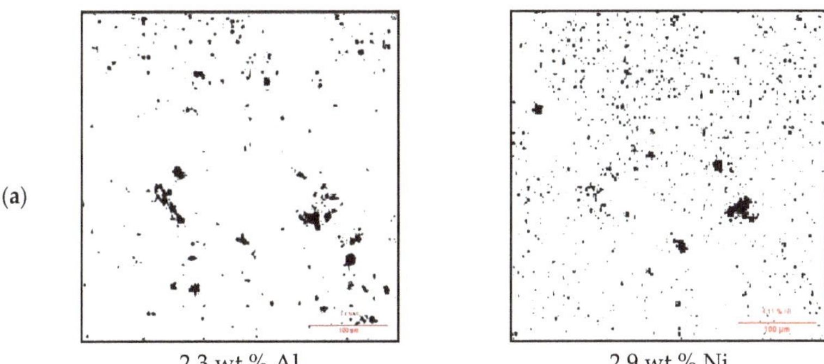

(a)

2.3 wt.% Al   2.9 wt.% Ni

Figure 10. *Cont.*

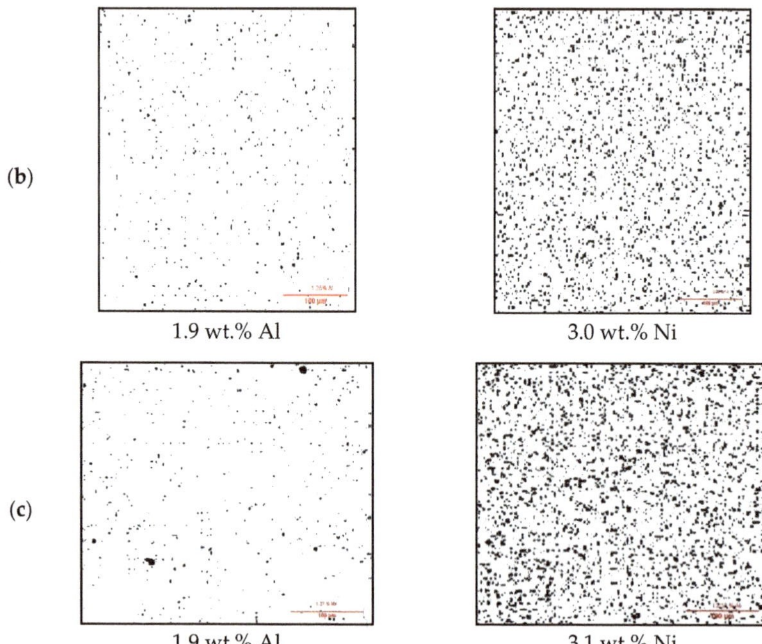

Figure 10. Elemental mapping for (**a**) As-received Cu CC from used batteries, (**b**) Cu CC from recycled batteries after melting, and (**c**) Cu CC from recycled batteries after reuse and cycling.

3.3. Cycling Performance of Recycled CCs

Results for testing half-cell anodes at a C-rate of C/5 using recycled Cu CCs are shown in Figure 11a. After five cycles, the capacity is 312 mAh/g while the capacity after ten cycles is 250 mAh/g. Capacity continues to drop until it reaches 50% after 25 cycles. Contrastingly for testing at a C-rate of C/10, shown in Figure 11b, the initial capacity after ten cycles was 255 mAh/g but it dropped to 220 mAh/g after fifteen cycles. The capacity reaches 67% of its initial value after 25 cycles measuring as 170 mAh/g.

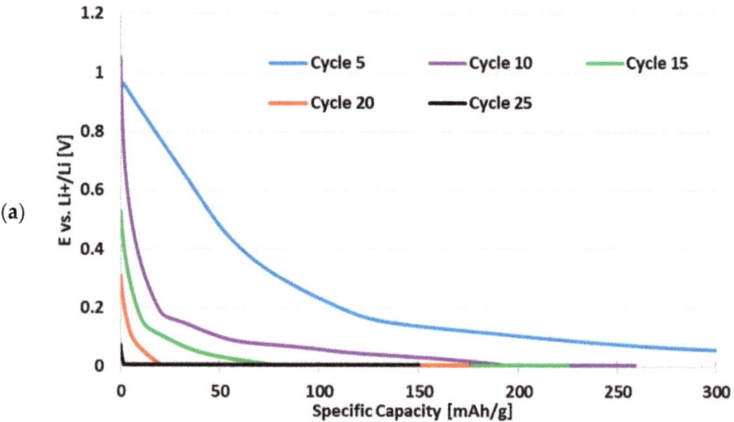

Figure 11. *Cont.*

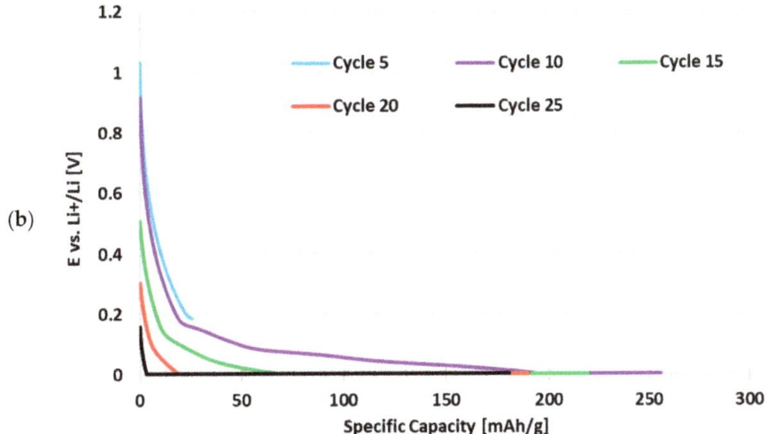

Figure 11. Discharge of anode half-cells with C-rates of (**a**) C/5, and (**b**) C/10.

Results for discharging cathode half-cells at C/5 C-rate are shown in Figure 12a. After 21 cycles, the capacity is 138 mAh/g. Capacity drops to 85 mAh/g after 52 cycles and finally reaches 39 mAh/g after 71 cycles. Moreover, Figure 12b includes results for discharging half-cell cathodes at C/10 C-rate. This showed an initial capacity of 157 mAh/g which consistently drops every 25 cycles until reaching a capacity of 130 mAh/g after 50 cycles. After 70 cycles, the capacity is 90 mAh/g which is 57% of the initial capacity. For all tests, specific capacity was normalized by the mass of electrode active material.

The capacity retention of the cathode half-cell tested at both C-rates is shown in Figure 13a along with a comparison to cathodes using fresh Al CC. For C/5, the capacity retention after 50 cycles is 84% while it is 83% for C/10. In contrast, the fresh CC cathode shows a capacity retention of 96%. When compared to the fresh CC cathode, capacity is lower; however, the cathode half-cells still retain more than 80% of the initial capacity. Capacity retention results for anode half-cells are shown in Figure 13b. The two test conditions of C/5 and C/10 do not show any significant difference in capacity retention

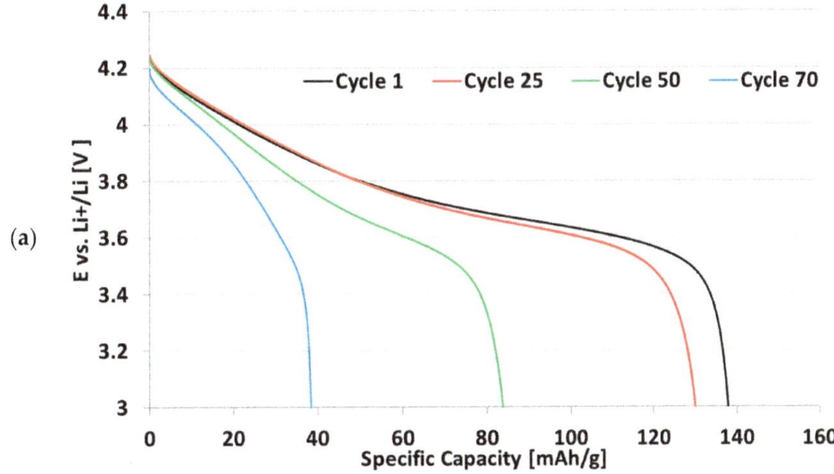

Figure 12. *Cont.*

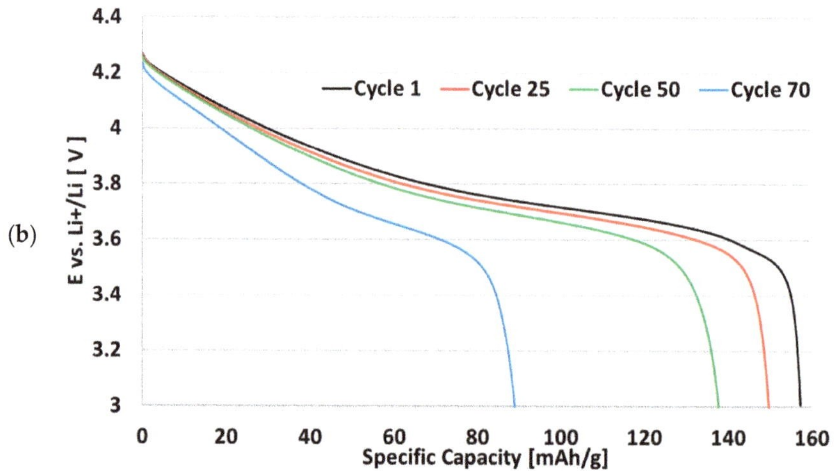

Figure 12. Discharge of cathode half-cells with C-rates of (**a**) C/5, and (**b**) C/10.

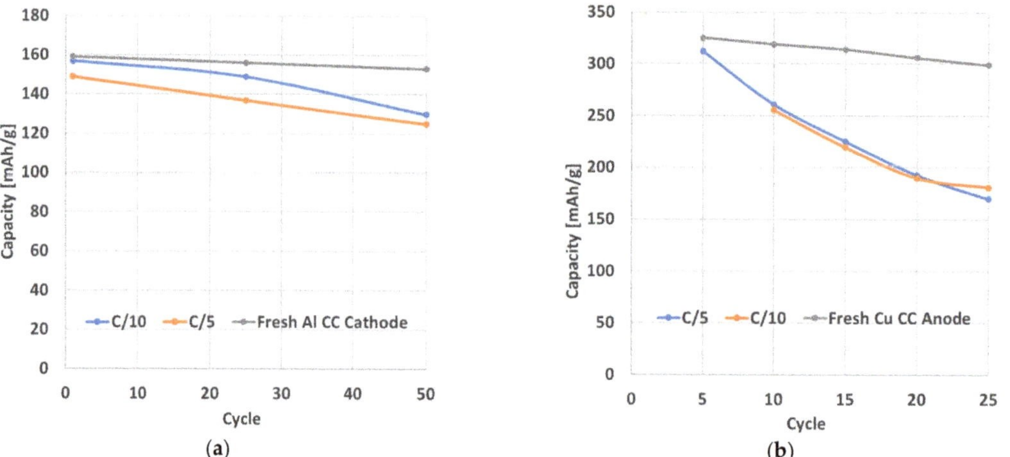

Figure 13. The capacity retention of (**a**) cathode half-cells over 50 cycles and (**b**) anode half-cells over 25 cycles compared to cathodes and anodes made from fresh Al and Cu CC.

4. Conclusions

Because of their high purity, Al and Cu CCs are among those high-value materials in LIBs manufacturing; hence, recycling and returning them to the LIB's manufacturing line is attracting great interest, especially in the EV market. Therefore, in this paper, we investigated the feasibility of recycling, purification, and reusing them in new LIBs. For this purpose, Al and Cu CCs from retired LIBs were recycled, purified, and reused in fresh LIBs. A brief description of the phases of the studies performed in this paper is as follows. In the first phase, shredded LIB cells were separated to obtain the components containing Al and Cu CCs. After separation, the ultrasonic cleaning of the CCs was investigated using several solvents. This includes an in-depth assessment of impurities detectable on or even below the surface (in depth) of the CC. The second phase of this study included recycling the cleaned Al and Cu CCs via melting and molding. Another assessment of impurities was conducted to show the effectiveness of the recycling procedure. Finally, the third phase of the investigation included constructing and testing anode and cathode half-cells at C-rates of C/5 and C/10 at room temperature. These cells used the same material from the two previous phases. It was found that ultrasonic cleaning of both Al and Cu CCs with ethanol was the most effective method, reducing a quarter of surface impurities. Material cleaned using this method and then recycled showed very low impurity indicating a highly effective process for purifying the surface of CCs. Battery cell testing of the aforementioned material showed an expected increase in impurity, especially when nickel is concerned. This is due to the CC degradation which naturally occurs during battery operation and may indicate loss of active cathode material. Battery testing results of both cathode and anode half-cells presented a decrease in capacity retention over time. However, after 50 cycles of testing at either C-rate, cathode half-cells capacity retention remained above 80%. Battery cell performance testing showed that both anode and cathode half-cells reach below their initial capacities after relatively short cycling periods. Although the cleaning and melting procedures significantly improved the purification level of both Al and Cu, it can be seen from battery cell performance results that CCs recycled using such a method may not be suitable for reuse in new batteries, especially for the Cu CC. Thus, a more sophisticated purification method should be adopted, which may increase the cost of recycling, or the recycled Al and Cu from retired LIBs should be repurposed in other products/applications where less material purity is acceptable.

Author Contributions: Conceptualization, S.F.; Methodology, S.F.; Formal analysis, H.K., E.H., D.F., M.A. and S.F.; Investigation, H.K., E.H., D.F., M.A. and R.F.; Resources, R.F. and S.F.; Data curation, H.K., E.H. and D.F.; Writing—original draft, H.K., E.H. and D.F.; Writing—review & editing, E.H., D.F., M.A. and S.F.; Visualization, E.H., H.K. and M.A.; Supervision, M.A., R.F. and S.F.; Project administration, S.F. and R.F. All authors have read and agreed to the published version of the manuscript.

Funding: The funding was provided internally by the Schaeffler Company and the University of Akron.

Data Availability Statement: Not applicable.

Acknowledgments: The support of the University of Akron and the Schaeffler Company is appreciated.

Conflicts of Interest: The authors declare no conflict of interest.

References

1. Bibra, E.; Connelly, E.; Gorner, M.; Paoli, L. Global EV Outlook 2021. International Energy Agency. Available online: https://iea.blob.core.windows.net/assets/ed5f4484-f556-4110-8c5c-4ede8bcba637/GlobalEVOutlook2021.pdf (accessed on 20 February 2022).
2. Huang, B.; Pan, Z.; Su, X.; An, L. Recycling of lithium-ion batteries: Recent advances and perspectives. *J. Power Sources* **2018**, *399*, 274–286. [CrossRef]
3. Kong, L.; Li, C.; Jiang, J.; Pecht, M.G. Li-Ion Battery Fire Hazards and Safety Strategies. *Energies* **2018**, *11*, 2191. [CrossRef]
4. Winslow, K.M.; Laux, T.G.; Townsend, A. Review on the growing concern and potential management strategies of waste lithium-ion batteries Resources. *Conserv. Recycl.* **2018**, *129*, 263–277. [CrossRef]
5. Chang, T.; You, S.; Yu, B.; Yao, K. A material flow of lithium batteries in Taiwan. *J. Hazard. Mater.* **2009**, *163*, 910–915. [CrossRef] [PubMed]
6. Melchor-Martínez, E.M.; Macias-Garbett, R.; Malacara-Becerra, A.; Iqbal, H.M.; Sosa-Hernández, J.E.; Parra-Saldívar, R. Environmental impact of emerging contaminants from battery waste: A mini review. *Case Stud. Chem. Environ. Eng.* **2021**, *3*, 100104. [CrossRef]
7. Song, J.; Yan, W.; Cao, H.; Song, Q.; Ding, H.; Lv, Z.; Zhang, Y.; Sun, Z. Material flow analysis on critical raw materials of lithium-ion batteries in China. *J. Clean. Prod.* **2019**, *215*, 570–581. [CrossRef]
8. Heelan, J.; Gratz, E.; Zheng, Z.; Wang, Q.; Chen, M.; Apelian, D.; Wang, Y. Current and Prospective Li-Ion Battery Recycling and Recovery Processes. *JOM* **2016**, *68*, 2632–2638. [CrossRef]
9. Sommerville, R.; Zhu, P.; Rajaeifar, M.A.; Heidrich, O.; Goodship, V.; Kendrick, E. A qualitative assessment of lithium ion battery recycling processes. *Resour. Conserv. Recycl.* **2021**, *165*, 105219. [CrossRef]
10. Kim, S.; Bang, J.; Yoo, J.; Shin, Y.; Bae, J.; Jeong, J.; Kim, K.; Dong, P.; Kwon, K. A comprehensive review on the pretreatment process in lithium-ion battery recycling. *J. Clean. Prod.* **2021**, *294*, 126329. [CrossRef]
11. Wang, X.; Gaustad, G.; Babbitt, C.W.; Richa, K. Economies of scale for future lithium-ion battery recycling infrastructure. *Resour. Conserv. Recycl.* **2014**, *83*, 53–62. [CrossRef]
12. Werner, D.; Peuker, U.A.; Mütze, T. Recycling Chain for Spent Lithium-Ion Batteries. *Metals* **2020**, *10*, 316. [CrossRef]

13. Lander, L.; Cleaver, T.; Rajaeifar, M.A.; Nguyen-Tien, V.; Elliott, R.J.; Heidrich, O.; Kendrick, E.; Edge, J.S.; Offer, G. Financial viability of electric vehicle lithium-ion battery recycling. *iScience* **2021**, *24*, 102787. [CrossRef] [PubMed]

14. Lai, X.; Huang, Y.; Gu, H.; Deng, C.; Han, X.; Feng, X.; Zheng, Y. Turning waste into wealth: A systematic review on echelon utilization and material recycling of retired lithium-ion batteries. *Energy Storage Mater.* **2021**, *40*, 96–123. [CrossRef]

15. Alhadri, M.; Zakri, W.; Farhad, S. Study on Integration of Retired Lithium-Ion Battery with Photovoltaic for Net-Zero Electricity Residential Homes. *J. Sol. Energy Eng.* **2022**, *145*, 21011. [CrossRef]

16. Alhadri, M.; Zakri, W.; Esmaeeli, R.; Farhad, S. A Study on Degradation of Lithium-Ion Batteries for In Aircraft Applications. In Proceedings of the ASME 2021, International Mechanical Engineering Congress and Exposition (IMECE), Virtual, 1–4 November 2021. [CrossRef]

17. Ayoola, O.M.; Buldum, A.; Farhad, S.; Ojo, S.A. A Review on the Molecular Modeling of Argyrodite Electrolytes for All-Solid-State Lithium Batteries. *Energies* **2022**, *15*, 7288. [CrossRef]

18. Mohammed, A.H.; Alhadri, M.; Zakri, W.; Aliniagerdroudbari, H.; Esmaeeli, R.; Hashemi, S.R.; Nadkarni, G.; Farhad, S. *Design and Comparison of Cooling Plates for a Prismatic Lithium-ion Battery for Electrified Vehicles*; SAE International Technical Paper; SAE International: Warrendale, PA, USA, 2018. [CrossRef]

19. Modjtahedi, A.; Amirfazli, A.; Farhad, S. Low catalyst loaded ethanol gas fuel cell sensor. *Sens. Actuators B Chem.* **2016**, *234*, 70–79. [CrossRef]

20. Figgener, J.; Stenzel, P.; Kairies, K.-P.; Linßen, J.; Haberschusz, D.; Wessels, O.; Angenendt, G.; Robinius, M.; Stolten, D.; Sauer, D.U. The development of stationary battery storage systems in Germany—A market review. *J. Energy Storage* **2020**, *29*, 101153. [CrossRef]

21. Fu, R.; Remo, T.W.; Margolis, R.M. *2018 U.S. Utility-Scale Photovoltaics-Plus-Energy Storage System Costs Benchmark*; No. NREL/TP-6A20-71714; National Renewable Energy Lab. (NREL): Golden, CO, USA, 2018. [CrossRef]

22. Dozein, M.G.; Mancarella, P. Frequency Response Capabilities of Utility-scale Battery Energy Storage Systems, with Application to the August 2018 Separation Event in Australia. In Proceedings of the 2019 9th International Conference on Power and Energy Systems (ICPES), Perth, Australia, 10–12 December 2019; pp. 1–6. [CrossRef]

23. Jiang, Y.; Jiang, J.; Zhang, C.; Zhang, W.; Gao, Y.; Li, N. State of health estimation of second-life LiFePO4 batteries for energy storage applications. *J. Clean. Prod.* **2018**, *205*, 754–762. [CrossRef]

24. Alkhalidi, A.; Alrousan, T.; Ishbeytah, M.; Abdelkareem, M.A.; Olabi, A. Recommendations for energy storage compartment used in renewable energy project. *Int. J. Thermofluids* **2022**, *15*, 100182. [CrossRef]

25. Bobba, S.; Mathieux, F.; Blengini, G.A. How will second-use of batteries affect stocks and flows in the EU? A model for traction Li-ion batteries. *Resour. Conserv. Recycl.* **2019**, *145*, 279–291. [CrossRef]

26. Abas, A.E.P.; Yong, J.; Mahlia, T.M.I.; Hannan, M.A. Techno-Economic Analysis and Environmental Impact of Electric Vehicle. *IEEE Access* **2019**, *7*, 98565–98578. [CrossRef]

27. Shafique, M.; Luo, X. Environmental life cycle assessment of battery electric vehicles from the current and future energy mix perspective. *J. Environ. Manag.* **2022**, *303*, 114050. [CrossRef] [PubMed]

28. Franzò, S.; Nasca, A. The environmental impact of electric vehicles: A novel life cycle-based evaluation framework and its applications to multi-country scenarios. *J. Clean. Prod.* **2021**, *315*, 128005. [CrossRef]

29. Flexer, V.; Baspineiro, C.F.; Galli, C.I. Lithium recovery from brines: A vital raw material for green energies with a potential environmental impact in its mining and processing. *Sci. Total. Environ.* **2018**, *639*, 1188–1204. [CrossRef] [PubMed]

30. Zhang, T.; Bai, Y.; Shen, X.; Zhai, Y.; Ji, C.; Ma, X.; Hong, J. Cradle-to-gate life cycle assessment of cobalt sulfate production derived from a nickel–copper–cobalt mine in China. *Int. J. Life Cycle Assess.* **2021**, *26*, 1198–1210. [CrossRef]

31. Choi, H.; Shin, J.; Woo, J. Effect of electricity generation mix on battery electric vehicle adoption and its environmental impact. *Energy Policy* **2018**, *121*, 13–24. [CrossRef]

32. Vieceli, N.; Casasola, R.; Lombardo, G.; Ebin, B.; Petranikova, M. Hydrometallurgical recycling of EV lithium-ion batteries: Effects of incineration on the leaching efficiency of metals using sulfuric acid. *Waste Manag.* **2021**, *125*, 192–203. [CrossRef]

33. Rajaeifar, M.A.; Raugei, M.; Steubing, B.; Hartwell, A.; Anderson, P.A.; Heidrich, O. Life cycle assessment of lithium-ion battery recycling using pyrometallurgical technologies. *J. Ind. Ecol.* **2021**, *25*, 1560–1571. [CrossRef]

34. Mohr, M.; Peters, J.F.; Baumann, M.; Weil, M. Toward a cell-chemistry specific life cycle assessment of lithium-ion battery recycling processes. *J. Ind. Ecol.* **2020**, *24*, 1310–1322. [CrossRef]

35. Yamada, M.; Watanabe, T.; Gunji, T.; Wu, J.; Matsumoto, F. Review of the Design of Current Collectors for Improving the Battery Performance in Lithium-Ion and Post-Lithium-Ion Batteries. *Electrochem* **2020**, *1*, 124–159. [CrossRef]

36. Zhu, P.; Gastol, D.; Marshall, J.; Sommerville, R.; Goodship, V.; Kendrick, E. A review of current collectors for lithium-ion batteries. *J. Power Sources* **2021**, *485*, 229321. [CrossRef]

37. Kim, S.W.; Cho, K.Y. Current Collectors for Flexible Lithium Ion Batteries: A Review of Materials. *J. Electrochem. Sci. Technol.* **2015**, *6*, 1–6. [CrossRef]

38. Pathak, R.; Chen, K.; Wu, F.; Mane, A.U.; Bugga, R.V.; Elam, J.W.; Qiao, Q.; Zhou, Y. Advanced strategies for the development of porous carbon as a Li host/current collector for lithium metal batteries. *Energy Storage Mater.* **2021**, *41*, 448–465. [CrossRef]

39. Noelle, D.J.; Wang, M.; Qiao, Y. Improved safety and mechanical characterizations of thick lithium-ion battery electrodes structured with porous metal current collectors. *J. Power Sources* **2018**, *399*, 125–132. [CrossRef]

40. Myung, S.-T.; Hitoshi, Y.; Sun, Y.-K. Electrochemical behavior and passivation of current collectors in lithium-ion batteries. *J. Mater. Chem.* **2011**, *21*, 9891–9911. [CrossRef]
41. Al-Shammari, H.; Farhad, S. Performance of Cathodes Fabricated from Mixture of Active Materials Obtained from Recycled Lithium-Ion Batteries. *Energies* **2022**, *15*, 410. [CrossRef]
42. Al-Shammari, H.; Farhad, S. Heavy liquids for rapid separation of cathode and anode active materials from recycled lithium-ion batteries. *Resour. Conserv. Recycl.* **2021**, *174*, 105749. [CrossRef]
43. Al-Shammari, H.; Farhad, S. Separating battery nano/microelectrode active materials with the physical method. In *Nanotechnology for Battery Recycling, Remanufacturing, and Reusing*; Farhad, S., Gupta, R.K., Yasin, G., Nguyen, T.A., Eds.; Elsevier: Amsterdam, The Netherlands, 2022; Chapter 13; pp. 263–286. [CrossRef]
44. Al-Shammari, H.; Farhad, S. Effects of imperfect separation of cathode active materials in recycling facilities on the performance of remanufactured lithium-ion batteries. In *Nanotechnology for Battery Recycling, Remanufacturing, and Reusing*; Farhad, S., Gupta, R.K., Yasin, G., Nguyen, T.A., Eds.; Elsevier: Amsterdam, The Netherlands, 2022; Chapter 21; pp. 445–453. [CrossRef]
45. Al-Shammari, H.; Farhad, S. *Regeneration of Cathode Mixture Active Materials Obtained from Recycled Lithium Ion Batteries*; SAE International Technical Paper; SAE International: Warrendale, PA, USA, 2020. [CrossRef]
46. Al-Shammari, H.; Esmaeeli, R.; Aliniagerdroudbari, H.; Alhadri, M.; Hashemi, S.R.; Zarrin, H.; Farhad, S. Recycling Lithium-Ion Battery: Mechanical Separation of Mixed Cathode Active Materials. In Proceedings of the ASME 2019, International Mechanical Engineering Congress and Exposition (IMECE), Salt Lake City, UT, USA, 11–14 November 2019. [CrossRef]
47. Farhad, S.; Gupta, R.K.; Yasin, G.; Nguyen, T.A. *Nanotechnology for Battery Recycling, Remanufacturing, and Reusing*; Elsevier: Amsterdam, The Netherlands, 2022; ISBN 978-032391134.
48. Kay, I.; Farhad, S.; Mahajan, A.; Esmaeeli, R.; Hashemi, S.R. Robotic Disassembly of Electric Vehicles' Battery Modules for Recycling. *Energies* **2022**, *15*, 4856. [CrossRef]
49. Kay, I.; Esmaeeli, R.; Hashemi, S.R.; Mahajan, A.; Farhad, S. Recycling Li-Ion Batteries: Robotic Disassembly of Electric Vehicle Battery Systems. In Proceedings of the ASME 2019, International Mechanical Engineering Congress and Exposition (IMECE), Salt Lake City, UT, USA, 11–14 November 2019. [CrossRef]

Article

Robotic Disassembly of Electric Vehicles' Battery Modules for Recycling

Ian Kay, Siamak Farhad *, Ajay Mahajan *, Roja Esmaeeli and Sayed Reza Hashemi

Department of Mechanical Engineering, College of Engineering and Polymer Science, The University of Akron, Akron, OH 44325, USA; ipk2@uakron.edu (I.K.); re25@uakron.edu (R.E.); sh184@uakron.edu (S.R.H.)
* Correspondence: sfarhad@uakron.edu (S.F.); majay@uakron.edu (A.M.)

Abstract: Manual disassembly of the lithium-ion battery (LIB) modules of electric vehicles (EVs) for recycling is time-consuming, expensive, and dangerous for technicians or workers. Dangers associated with high voltage and thermal runaway make a robotic system suitable for the automated or semi-automated disassembly of EV batteries. In this paper, we explore battery disassembly using industrial robots. To understand the disassembly process, human workers were monitored, and the operations were analyzed and broken down into gripping and cutting operations. These operations were selected for automation, and path planning was performed offline. For the gripper, a linear quadratic regulator (LQR) control system was implemented. A system identification method was also implemented in the form of a batch least squares estimator to form the state space representation of the planar linkages used in the control strategy of the gripper. A high-speed rotary cut-off wheel was adapted for the robot to perform precise cutting at various points in the battery module case. The simulation results were used to program an industrial robot for experimental validation. The precision of the rotary cutter allowed for a more direct disassembly method as opposed to the standard manual method. It was shown that the robot was almost twice as fast in cutting but slower in pick and place operations. It has been shown that the best option for disassembly of a LIB pack is a human–robot collaboration, where the robot could make efficient cuts on the battery pack and the technician could quickly sort the battery components and remove connectors or fasteners with which the robot would struggle. This collaboration also reduces the danger encountered by the technician because the risk of shorting battery cells while cutting would be eliminated, but the time efficiency would be significantly improved. This paper demonstrates that a robot offers both safety and time improvements to the current manual disassembly process for EV LIBs.

Keywords: lithium-ion battery; recycling; robots; automation; electric/hybrid vehicles

Citation: Kay, I.; Farhad, S.; Mahajan, A.; Esmaeeli, R.; Hashemi, S.R. Robotic Disassembly of Electric Vehicles' Battery Modules for Recycling. *Energies* **2022**, *15*, 4856. https://doi.org/10.3390/en15134856

Academic Editor: Cai Shen

Received: 9 June 2022
Accepted: 29 June 2022
Published: 2 July 2022

Publisher's Note: MDPI stays neutral with regard to jurisdictional claims in published maps and institutional affiliations.

1. Introduction

Robotic systems have been widespread in usage for the assembly of lithium-ion batteries (LIBs) for many years, but their use for disassembly has only recently been considered as hybrid electric vehicles (HEVs) and electric vehicles (EVs) become more prevalent. As the adoption of EVs and HEVs gains traction, the value of the active materials, such as lithium cobalt oxide ($LiCoO_2$) and lithium nickel manganese cobalt oxide ($LiNiMnCoO_2$), within the batteries is lost if they are not recycled [1–3]. To prepare for the end of life (EOL) of these battery systems, the recycling process for LIBs is being researched to reduce not only the environmental impact of landfilling LIBs but also to recover the valuable materials contained within the LIBs [4–6]. It has been shown that the robotic disassembly of an HEV battery system is possible [7], but more work is required in this field for proof of concept.

Most LIB recycling is done through pyrometallurgical [8], hydrometallurgical [9,10], and physical/direct processes [11]. These methods may rely on the disassembly and shredding of battery cells/modules/packs before the process begins to reduce the contaminants created by the case and structural materials used in the construction of modules and packs. The disassembly process allows for the sorting of bulk materials, such as aluminum, structural plastics, printed circuit boards (PCBs) and integrated circuits (ICs), copper, and various other components [9]. In the physical/direct recycling process, once these materials are removed, either individual cells or modules can be put into a shredder to separate the materials contained within the cells. The shredded cells or modules then undergo a physical separation process and are submerged in an alkali solution of lithium brine. This process allows the various materials to be sorted via physical and chemical means, increasing the purity of the filter/powder cake [12] and making the various by-products suitable for secondary processing.

Currently, more than 95% of LIBs are landfilled each year [13]. In addition, the total number of unrecycled batteries are increasing because of their adoption in EV applications [13]. Some of the major challenges in LIB recycling are due to the economics of the recycling process. Currently, the costs associated with the recycling of LIBs may be enough to make recycling unsustainable without subsidies or mandates due to the recovered value of materials being less than the required resources. One cost that significantly impacts the recycling process is the labor associated with the disassembly of packs and modules. This step in the recycling process is both time consuming and hazardous to the technicians/laborers. The manual disassembly process requires a technician/laborer to operate hand tools to remove fasteners, connectors, wires, and cut tabs, which can be difficult due to the pack layout and location of various hardware components within the pack assembly. In addition, the technician/laborer must wear personal protection equipment (PPE) and handle the tools with extreme care, as any mistake in the disassembly process could present hazards to the technician/laborer in the form of electrocution or explosion. Hazards associated with the disassembly process are a result of the high voltage in the battery system if it is not properly discharged and thermal instability if the individual cells are mechanically, electrically, or thermally compromised. The hazards can be mitigated through the proper pre-disassembly procedures, but caution should always be taken when handling the packs for disassembly.

To reduce the economic burden associated with the disassembly process, a robotic method for automating or semi-automating the disassembly process was envisioned and tested in disassembling a representative battery module. This process was broken down into the following steps: process identification, suitable actions for robotic implementation, end-of-arm tooling design and control, development of a representative work piece, workspace layout, path planning, robotic modelling and simulation in MATLAB/Simulink, and experimental testing of the robotic system. The research focused on the implementation of a robot for disassembly, and the programming of the robot was done through direct inputs from the user with either the teach pendant [14] or manual positioning [14]. This methodology has the advantage that it can be done by technicians without prior knowledge since the path planning is situationally dependent. The human brain is quickly able to make decisions as to how a system needs to be assembled or disassembled, and this process is not easily automated due the highly complex nature of electro-mechanical systems. Research has been performed in the area of machine learning for the purpose of disassembly automation [14]. Oak Ridge National Laboratory has also started introducing an automated disassembly line to make battery recycling safer and faster [15]. Recently, some researchers reported studies on task planners for robot disassembly [16] and human–robot collaboration [17], but more research is required in this field.

In this study, a novel approach for disassembly of EV/HEV LIBs is presented based on the off-line simulation and path planning, as proposed to aid the implementation of the disassembly process. By using offline path planning, the precise control and tool paths of the robot can be predefined to maximize the efficiency of the required disassembly steps. In addition to this, both the robot dynamics and process time can be quantified to provide more insight into the state of the robot. By utilizing the CAD geometry models directly, offline simulation can be performed ahead of time, and allows the operator the ability to precisely define waypoints within the model that can allow for alternative disassembly techniques.

2. Disassembly Workflow

To determine what steps in the disassembly process would be suitable for a robot, the manual disassembly process was observed and documented in Figure 1. The process of manual disassembly for the pack involved the following steps: (1) removal of the pack cover, (2) disconnection and removal of the main wiring harness and primary battery management system (BMS), (3) disconnection and removal of the thermal management system plumbing, (4) removal of the module tie downs, and (4) removal of the battery modules.

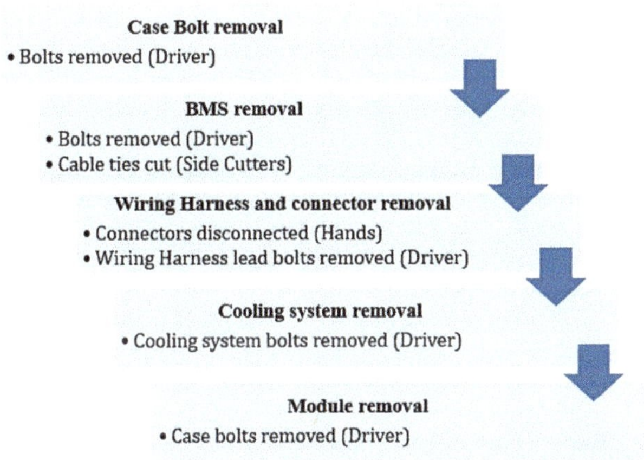

Figure 1. Disassembly process diagram of a battery pack by technician.

The disassembly of individual modules is comprised of the following: (1) the removal of the module BMS and main harness connector, (2) removal of the battery module frame bolts, (3) the removal of the module top cover, (4) the cutting of the first battery cell tabs, (5) the separation and removal of the first battery cell, and (6) the repetition of step 5 for the remaining cells. To describe the workflow of the disassembly process, one could maximize the use of the technician and robot by the following process, as depicted in Figure 2.

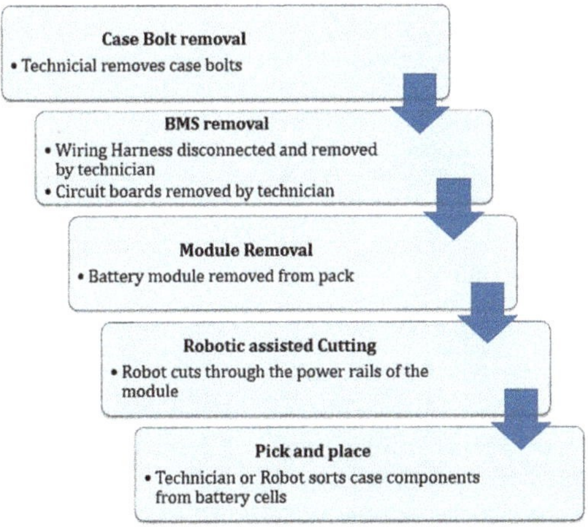

Figure 2. Process workflow as modified for a technician–robot collaboration.

3. Experimental Design for Human–Machine Collaboration

As the pack level disassembly of the case would require an industrial robot with a large reach as well as a high payload, it was decided that for the purposes of this experiment the scope of implementing a robot would be on the module disassembly level. A FANUC LR-Mate 200iD/4S, six degree of freedom (6DOF) industrial robotic system was used with a maximum payload of 4 kg and maximum reach of 550 mm. This robot was chosen due to its precision and small size, as well as having the ability to communicate with an end effector through the input/output port on the upper arm.

To prove the suitability of a robot for the disassembly of an EV/HEV battery system, the benchmark operations of cutting and gripping were selected for evaluation. Both the cutter and gripper are required for the disassembly process by the robot, and they are add-ins to the robot. Fine motor function, such as disconnection of electrical connectors, were decided to be outside the scope of this experiment, as the inherent level of complexity and dexterity needed would require more specialized end-of-arm tooling. A more generalized assessment of the robot's ability to perform gross motor function was chosen. The assessment was to be done for single end-effectors without accounting for the tool-change time, as a quick tool change and docking station could be implemented to automate the tool-change process. The total time for module disassembly would be evaluated based on the sum of the cutting and gripping operations, as well as any additional collaboration needed from a technician, such as fastener, connector, or harness removal. It is noted that the process analysis to find the total time of the disassembly process by robots is out of the scope of this paper.

End-of-arm tooling was a major focus of the experimental design, as the functional requirements defined in the robotic implementation stage required both highly precise motion and careful manipulation of components to prevent accidental shorting of the batteries or causing mechanical damage to the individual cells which could trigger a thermal event where explosion or fire could occur. Multiple modalities for the cutting tool were conceptualized, and a weighted decision matrix was used to select the best solution for implementation, design, and prototyping.

The first end effector to be evaluated was the cutting instrument, as this would be responsible for dismantling and separating the casing components from the individual battery cells. The cutting instrument would need to be compact, as the payload is limited to 4 kg, while maintaining the ability to cut through plastics and metal objects with both precision and speed. Cost was also considered a concern, as any additional cost of the robotic system would reduce the economic benefit. As listed in Table 1, the three modalities of cutting investigated were a high-speed rotary cutter, a reciprocating saw mechanism, and a laser-cutting instrument.

Table 1. Weighted decision matrix for three cutting methods.

Method	Cost (25)	Form Factor (5)	Heat Generation (15)	Vibration Amplitude (15)	Cut Precision (15)	Technical Implementation Difficulties (25)	Total (100)
Rotary cutter	25	5	10	10	13	25	88
Reciprocating cutter	20	4	15	5	5	15	64
Laser cutter	10	3	10	15	15	20	73

The high-speed rotary cutter was chosen due to the precision of cutting, compact size, and less technical difficulties, while still maintaining acceptable levels of heat generation and vibration transmission. Due to high-speed rotary cut-off tools being common, a Dremel 4000 was adapted to perform the cutting operations, as the motor and cutting speed were able to accommodate a wide range of materials and thicknesses. This was implemented, as shown in Figure 3a, and used in the offline simulation to represent the geometric properties.

Threads in the tool handle made cutting-tool integration with the robot secure and efficient. Since the state of the cutting tool only had one parameter that was controllable without modification of the tool, input/output states for the cutting tool can be controlled via a digital output as an on/off state from the upper arm in/out port communicating with a power relay to the tool.

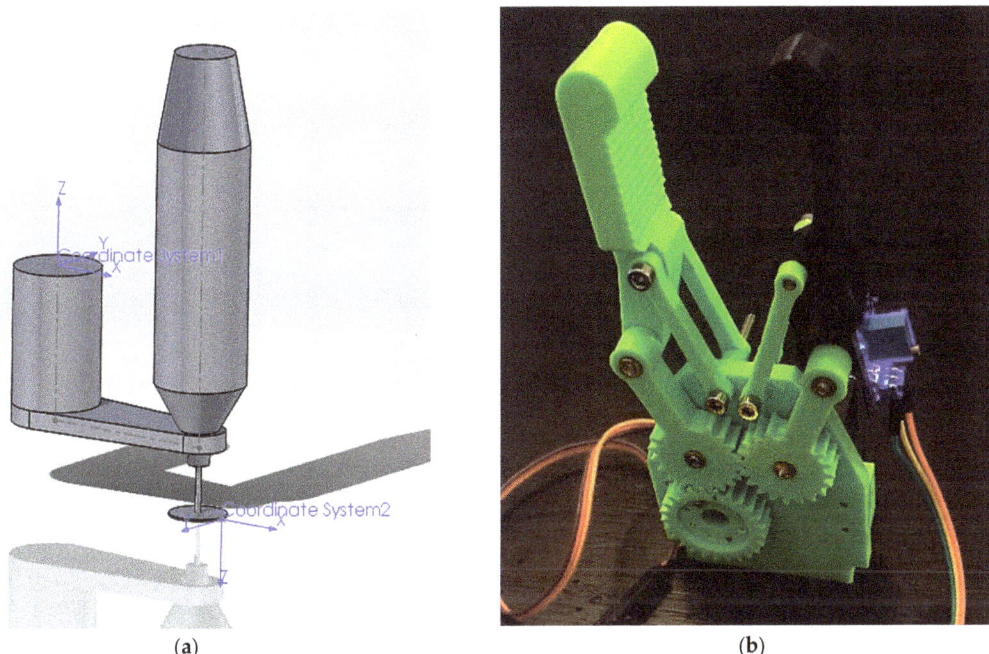

(a)   (b)

Figure 3. (**a**) The CAD model of the cutting tool, (**b**) the 3D-printed griper with servo actuator and gear-driven coupled planar linkage system with force feedback.

After determining how to implement the cutting-tool control and designing custom mounting for it, the gripper was designed and built, as shown in Figure 3b. Most robotic cells have gripper attachments, but few control the exact gripping pressure or force that grippers use to interface components. This is particularly important in the case of LIBs, as mechanical damage to the battery cell can cause internal short circuit, leading to thermal runaway [18,19] which is a major hazard presented by LIBs. For the gripper to not damage battery cells, force regulation became an important feature in the gripper design. The primary design for the gripper was based on a servo-actuated, gear-driven coupled planar linkage system with force feedback as the feature to be controlled. This was done through the implementation of a strain gauge into the gripper's finger to measure the strain of the finger under load.

The finger was modeled as a fixed-free cantilever beam, where the load was applied at the tip of the finger. Knowing the geometric dimensions and the material properties of the as printed PLA, the Young's modulus, the beams cross sectional properties, and the length of the beam were used to determine the bending stress at the base of the beam through the bending-stress equation in Equation (1).

$$\sigma = \frac{My}{I} = \frac{FLh}{\frac{1}{6}bh^3}.$$

(1)

The force was then calculated from Equation (2) to determine how the input mapped to the output of the system for setting up the error value.

$$F = \frac{6L}{E\varepsilon bh^2}.$$

(2)

The general design of the gripper was modeled as two gear-driven coupled four bar linkages with nearly parallel input links and four revolute joints to create a smooth motion. This allowed for a simplified actuation principle, as the system has only one degree of freedom being the input link angle if backlash and linkage compliance are assumed to be minimal. The general procedure was formulated similar to Ref. [20].

After determining how to observe the forces, a mathematical model for the system was created to implement a feedback control law for the system. This was derived through the use of Lagrange's equations. However, this yielded a sixth order nonlinear system of state equations for the planar linkage of the gripper. As a result of operating away from the nonlinear behavior, the system was approximated as a third order, linear, single input/single output (SISO) system, and the state space representation for the system dynamics was synthesized through the use of a least squares identification method. After synthesizing the state equations for use with the feedback error in the force, an infinite horizon linear quadratic regulator was implemented to control the system. For convenience and first try, the performance index, Q, and R matrices were selected as follows:

$$J = \int_{t_0}^{t_f} \left(x^T Q x + u^T R u \right) dt,$$

(3)

$$Q = \begin{bmatrix} 1 & 0 & 0 \\ 0 & 1 & 0 \\ 0 & 0 & 1 \end{bmatrix}, \ R = [1].$$

The optimal gains were calculated through the solution to the algebraic Riccati equation (ARE) Equation (4) and implemented using state variable feedback, as shown below in Equations (5) and (6).

$$A^T S + SA + Q - SBR^{-1}B^T S = 0,$$

(4)

$$K = R^{-1}B^T S,$$

(5)

$$u = -Kx.$$

(6)

This was enough for the case of contacting the work piece. However, the initial conditions for solving the differential equations were not sufficient for the case of non-contact; thus, creating a two-step control strategy was required. As such, the gripper was modeled as a finite state machine with four states during operation. Due to the initial conditions not being available until after contact had been made with the work piece, four states were executed before the linear-quadratic regulator (LQR) control sequence was used. The first state was an idle state where the gripper was left in the state at which it last operated. After receiving an input signal from the robot during the pick and place sequence, the servo on the gripper moves the gripper to a limit open angle and sends a ready status to the robot in the form of a digital output. Once the robot moves the gripper into place for gripping the work piece, the robot then sends a start signal to the gripper to initiate a constant advance. After the contact force is detected via exceeding the contact force threshold, the initial conditions for the state equations are defined, and the LQR control sequence begins until the drop signal is given by the robot. The gripper controller was then tested for performance and implemented on the microcontroller through the Arduino toolbox in MATLAB, as shown in Figure 4.

Due to hazards associated with disassembly of a real battery system, a representative battery module was created that mimicked the geometry, layout, and material composition of the battery module observed in the manual disassembly. The construction of the module consisted of a module frame, cell compartments that also functioned as cooling plates, top and bottom module covers, a front tab harness and connector, module frame bolts, and the module BMS. This was positioned on a flat surface within complete reach of the arm, where both cutting operations and pick and place could be completed. The representative module was made through the use of a 3D printer and assembled in a similar way to the actual module. The goal was to position the battery module in the workspace such that the tabs were accessible and that the individual cells could be lifted out from the top. Additionally, the module needed to remain in the same location during the cutting procedure; thus, the outline of the module location was marked, and the module was held in place with weights during the cutting operation.

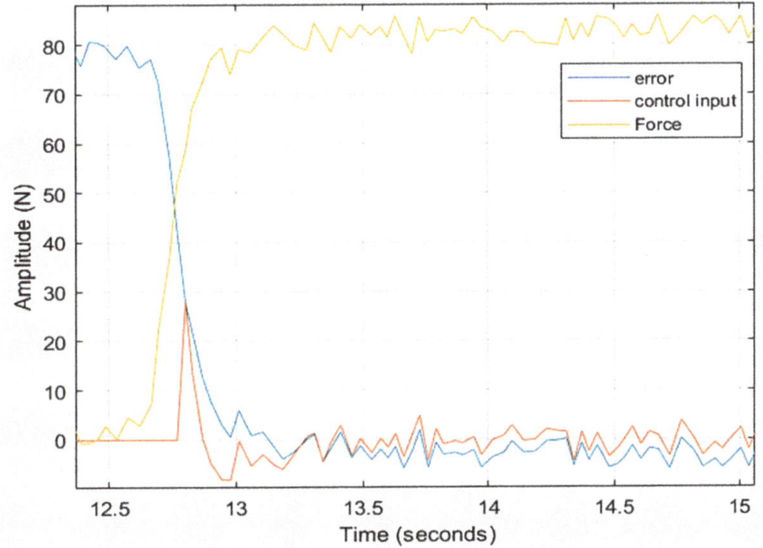

Figure 4. Gripper response under LQR control (results do not change if we increase the time).

To develop the kinematic model of the robot, the Denavit–Hartenberg (DH) representation of a kinematic linkage chain can be used to model the system in terms of the DH parameters, a_i, d_i, θ_i, α_i, as shown in Figure 5 [21].

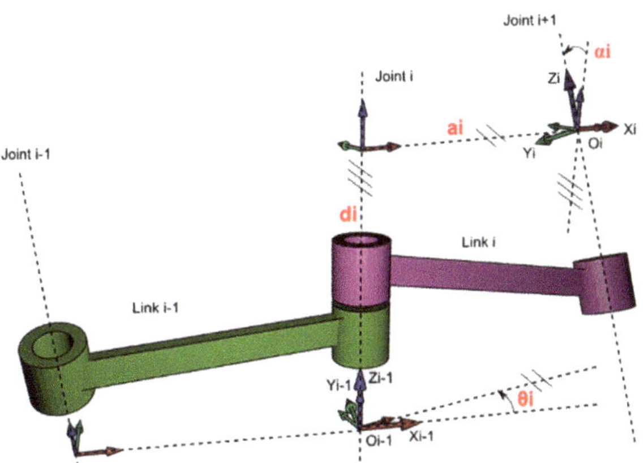

Figure 5. Parameter definitions of the Denavit–Hartenberg representation of a kinematic linkage chain.

For the case of the LR-Mate 200iD/4S, the DH parameters for the articulated robot are displayed in Table 2. J1-6 are the six links of the robot, and J7 is the end effector.

Table 2. DH parameters for the LR-Mate 200iD/4S [22].

Link	θ_i	d_i	a_i	α_i
J1	θ_1	0	160	$\pi/2$
J2	θ_2	0	330	0
J3	θ_3	0	260	$\pi/2$
J4	θ_4	290	0	$-\pi/2$
J5	θ_5	0	0	$\pi/2$
J6	θ_6	70	0	0
J7	0	107	100	$\pi/2$

Once the DH parameters are determined, the homogenous transformation matrix for each link can be computed as the product of the various rotation and translation matrices [21], as shown in Equation (7).

$$^{i-1}_{i}T(\alpha_i, \theta_i, a_i, d_i) = R_z(\theta_i) \times T_z(d_i) \times T_x(a_i) \times R_x(\alpha_i). \tag{7}$$

These equations are fundamental to both the forward and inverse kinematics solver for determining or solving the position and pose of the robot in space. Forward kinematics uses joint states to calculate a position given a specified pose of the robot, whereas inverse kinematics solve for the joint states of a given end effector position and pose. Solving the inverse kinematic equations is more difficult, as there can be multiple solutions or no solutions to a given end effector position.

Once the DH parameters are known along with all the masses and lengths of the links and end effector, one can use the following equation to solve for the torques required to move the robot joints to a desired location, and from there along a desired trajectory. This equation is often called the robot equation or the dynamic model for a robot.

$$\tau = D(q)\ddot{q} + h(q, \dot{q}) + C(q), \tag{8}$$

where τ is the joint torques, D is the inertia matrix for all the links, h is Coriolis and centrifugal terms, and C is the gravity terms.

MathWorks' Robotics System Toolbox and Simscape Multibody Dynamics software packages were used in the MATLAB/Simulink environments to solve these equations. Both the forward and inverse kinematics/dynamics are simplified into blocks for use in solving the equations. There is a way to link CAD models directly into Simscape Multibody in the form of a Unified Robot Description Format (urdf) file. To import the robot model, the CAD model must be assembled, and the various coordinate systems, joint axes, inertia properties, and joint limits need to be defined during the setup of the urdf file. Once the urdf file is setup, it can be imported into MATLAB/Simulink, and motion analysis and offline path planning can be performed. Once the path planning was completed, simulation (Figure 6a) was performed in MATLAB/Simulink to evaluate the motion of various joint states. The waypoints as well as pose information (Figure 6b) were imported into MATLAB and fed into the inverse kinematics block in Simulink.

MATLAB utilizes the Levenberg–Marquardt algorithm (LMA) to solve the nonlinear least squares problem for the inverse kinematics. This method utilizes the minimization of the error in the joint configuration to that of the desired configuration to solve for the closest robot input joint angles to reach a target point [20,23]. This method is very robust and capable of solving complex motion tasks but can be slow to perform in real time. The Simulink block diagram for path definition and inverse kinematic is shown in Figure 7.

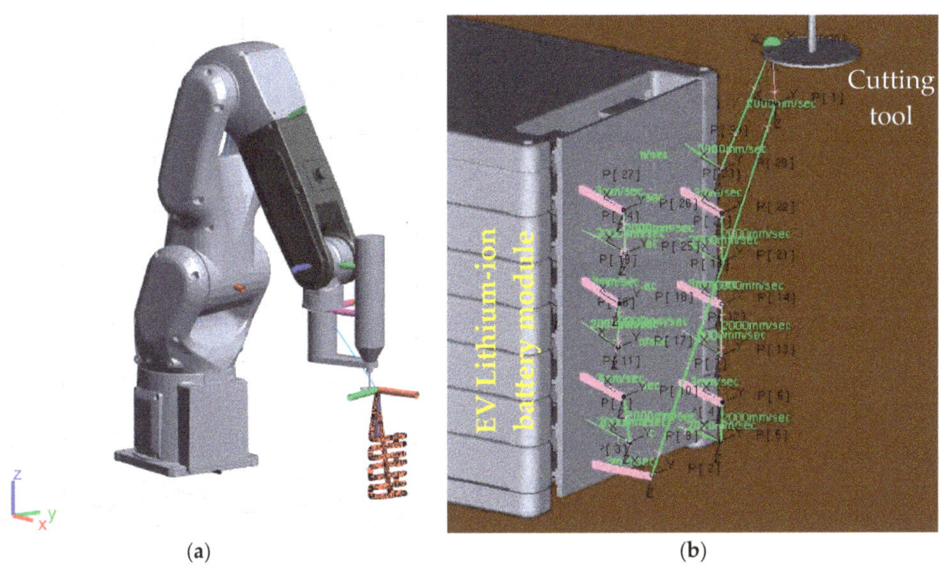

(**a**)   (**b**)

Figure 6. (**a**) Robot motion simulation to evaluate the motion of various joint states, (**b**) cutting path for a retired EV battery module to make it ready for recycling.

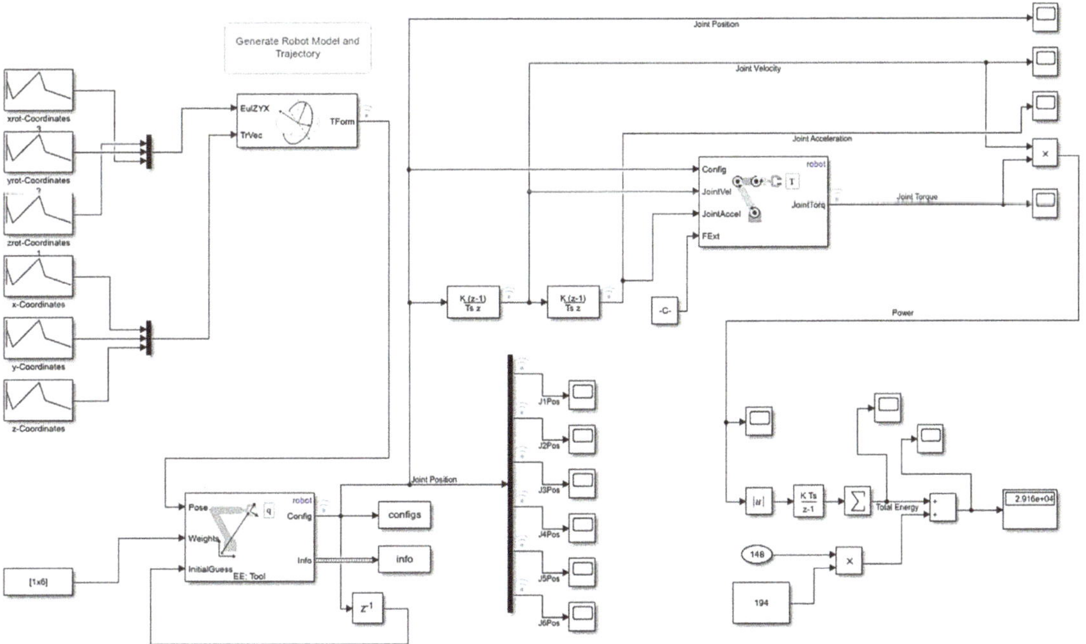

Figure 7. Simulink block diagram for path definition and inverse kinematic.

4. Results and Discussion

The velocity, acceleration, and torque were obtained from the offline simulation of cutting an EV battery module and then plotted for each robot joint, as demonstrated in Figure 8, to study the involvement of each joint in the process. As seen in this figure, the angular velocities for joint 2 and joint 5 were highest, and the angular acceleration for joint 5 and joint 7 were highest at the beginning and end of the process, as compared to other joints. The angular velocity and angular acceleration of all joints are fluctuating at the middle of the process; however, their values are not significantly greater than zero. The highest torque also belongs to joint 7, followed by joints 3 and 5. The torque of other joints remains close to zero during the process. As a remarkable result, it was revealed that all velocities, accelerations, and torques can be easily handled by the robot, as shown in the Figure 8. The error of values in Figure 8 depends on the accuracy of the model and the model input parameters. Our estimation is that the error of the Simulink simulation for the velocity, acceleration, and torque is less than 2%.

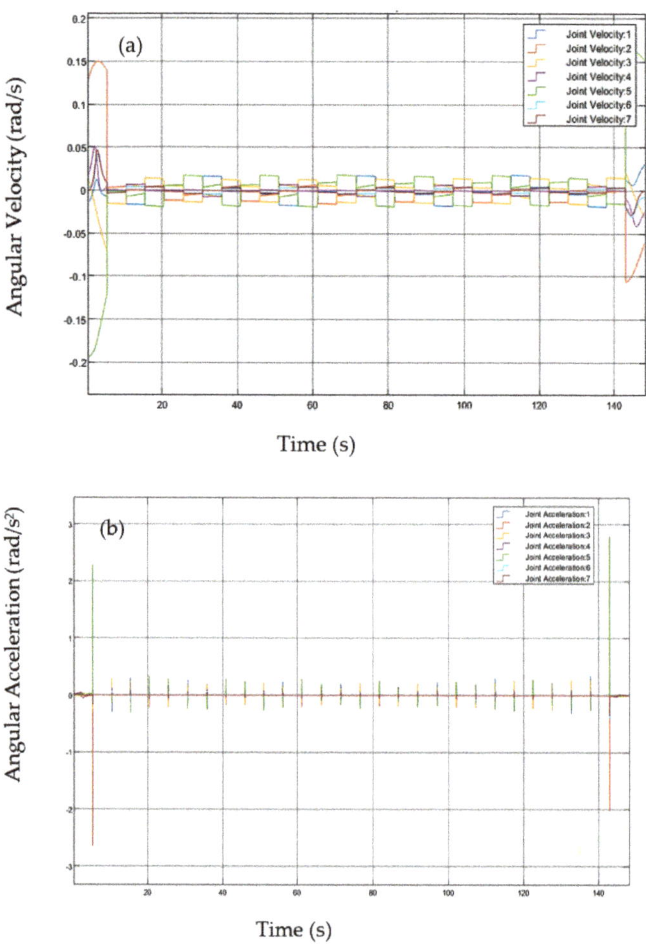

Figure 8. *Cont.*

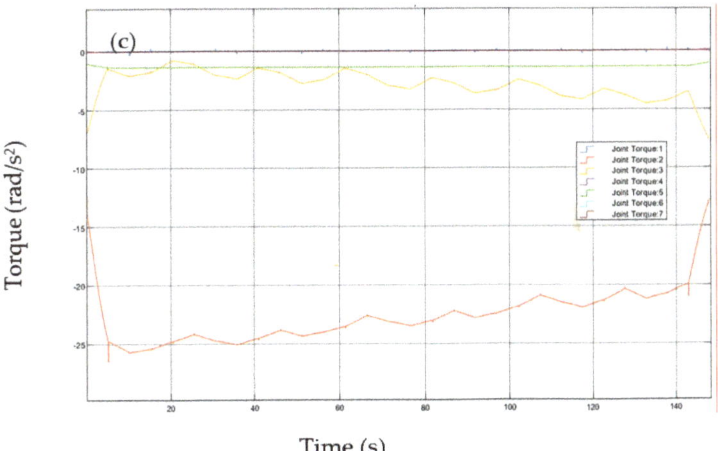

Figure 8. The results of Simulink simulation: (**a**) joint angular velocity vs. time, (**b**) joint acceleration vs. time, (**c**) joint torque vs. time.

After the offline simulation was performed, the robot program was uploaded to the teach pendant to begin making the initial cuts experimentally. The cutting path was the first item to be tested in the experiment. The quality of the cut was based on two performance metrics:

1. accuracy of the cut and
2. disruption of the work piece within the workspace.

All battery tabs in the EV battery modules (the LIB cells are the pouch type) were successfully cut by the robot (See Figure 9). The cutting tool placed the cut with both accuracy and speed, performing the operation in 112 s (1:52 min). A technician completed this operation in 220 s (3:40 min) in the LIB recycling plant, which is 108 s (1:48 min) longer than the robot. This almost a 2X gain in speed efficiency. Thus, the disassembly process may be two times faster when a robot is incorporated into the process. However, more detailed time analysis is required for the final answer.

However, more importantly, the cutting path developed for the robot was more efficient, as it was a continuous cutting process and did not require the tool to be changed. Pick-and-place operations proved to be slightly lengthier and less successful in both time and performance compared to the technician. Therefore, the best option for disassembly of a LIB pack/module would be a human and robot collaboration, where the robot could make efficient cuts on the battery pack, and the technician/laborer could quickly sort the battery components and remove connectors or fasteners with which the robot would struggle. This is based on the results of the experiment, as the technician is capable of sorting through the components of the battery module 18 s faster than the robot and with higher rates of success.

The readers are referred to Ref. [24], the publication of the authors at ASME IMECE-2019 for more information on the design aspect of this project. For more information about recycling lithium-ion batteries, the readers are recommended to look at Refs. [25–27].

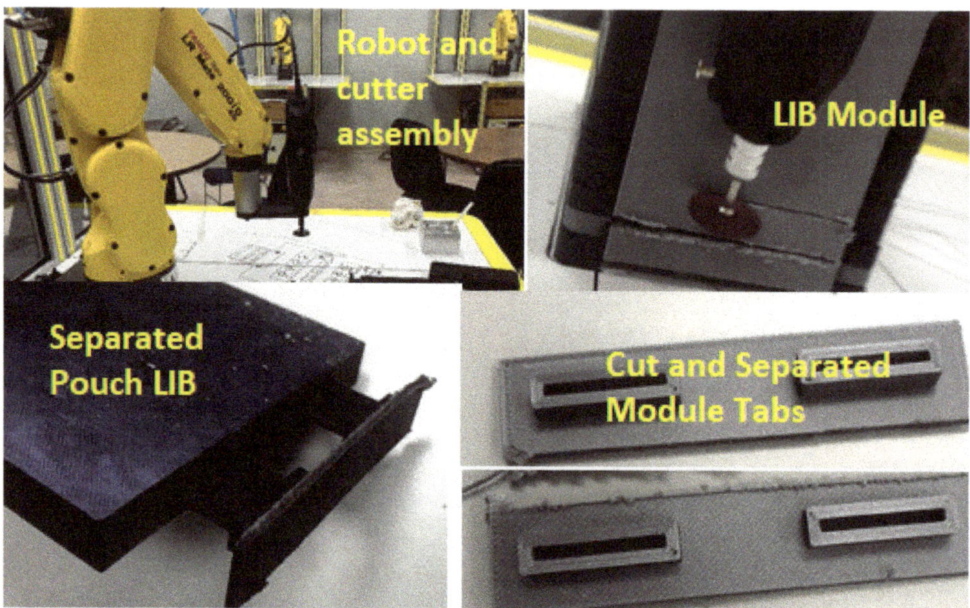

Figure 9. The results of the implementation of the work experimentally.

5. Conclusions

In this study, a method for the robotic disassembly of an electric vehicle battery module was proposed and tested. Offline simulation was used to perform path planning for the disassembly of the battery module, and custom end-of-arm tooling was designed. The cutting tool demonstrated improvements in effectiveness over a purely technician-based disassembly due to the precision of the cuts made by the robotic arm. The cutting speed and direction did have a significant influence on the cut quality, as too high of a speed would increase the temperature, and too low of a speed would stall the motor. In the experimental testing of the cutter, the robotic arm was extremely capable in making the required cuts, and the only instances it failed were when the mock battery module was not properly secured to the workspace. A way of controlling the gripper to utilize force regulation was designed using system identification techniques in the form of a batch least squares estimator. This was then utilized to form a state space representation for the gripper, and a linear quadratic regulator was implemented to determine the optimal gains required to control the system. Despite the development of a customized gripper under LQR control, the pick-and-place operations proved to be more of a challenge, as the control sequence was slow to sense contact with the part, and the geometry of the gripping tool was not able to securely hold on to the battery cell. This could be overcome through redesigning the gripper fingers, choosing more rigid construction materials, and developing a faster sensing method for contact detection. As a remarkable result, it can be concluded that the best option for disassembly of a LIB pack would be human and robot collaboration, where the robot could make efficient cuts on the battery pack, and the technician could quickly sort the battery components and remove connectors or fasteners with which the robot would struggle. This collaboration also reduces the danger encountered by the technician because the risk of shorting battery cells while cutting would be eliminated, and sharing in the labor contribution reduces the required man hours for disassembling an EV battery module or pack. The focus of this study was the technical aspects of the disassembly of EV/HEV LIBs by industry robots. The continuation of this work can be an economic analysis and studying the time saving in the disassembly process when the human–robot collaboration approach is chosen.

Author Contributions: Data curation, I.K.; Formal analysis, I.K., S.F., A.M., R.E. and S.R.H.; Funding acquisition, S.F.; Investigation, I.K. and S.F.; Methodology, I.K. and S.F.; Resources, S.F.; Software, S.F.; Supervision, S.F. and A.M.; Validation, I.K.; Writing—original draft, I.K., R.E. and S.R.H.; Writing—review & editing, S.F. and A.M. All authors have read and agreed to the published version of the manuscript.

Funding: This research received no external funding.

Acknowledgments: This work would not have been possible without the help of Don Goode and Ed Green from Retriev Technologies, the University of Akron for providing the necessary software and hardware resources and financial support, Ohio Department of Higher Education, and the RAPIDS program for funding the robot cells. A part of this paper has been presented at the ASME IMECE conference in 2019, Paper No: IMECE2019-11949, entitled "Robotic Disassembly of Electric Vehicle Battery Systems" by the authors.

Conflicts of Interest: The authors declare no conflict of interest.

Nomenclature

A	A matrix
b	width of cantilever beam (mm)
F	applied force (N)
B	B matrix
E	Young's modulus (GPa)
h	thickness of cantilever beam (mm)
I	second moment of area (mm^4)
M	bending moment (N.m)
l	length of beam (mm)
ε	strain
a_i	length of common normal (m)
d_i	offset along z-axis to common normal (m)
θ_i	angle about previous z (degrees)
α_i	angle about common normal (degrees)
J	performance index
x	generalized coordinate (mm/degrees)
Q	weighting matrix on the position
R	weighting matrix on the control input
u	control input
K	optimal gain
S	variable used for the ARE
T	homogenous transformation matrix
$R_z(\theta_i)$	rotation matrix about z axis
$T_z(d_i)$	translation matrix along z
$R_x(\alpha_i)$	rotation matrix about x
$T_x(a_i)$	translation matrix along x

Abbreviation

BMS	battery management system
EOL	end of life
HEV	hybrid electric vehicles
IC	integrated circuits
LQR	linear quadratic regulator
LIB	lithium-ion battery
PCB	printed circuit board

References

1. Wang, L.; Wang, X.; Yang, W. Optimal design of electric vehicle battery recycling network—From the perspective of electric vehicle manufacturers. *Appl. Energy* **2020**, *275*, 115328. [CrossRef]

2. Zhang, L.; Li, L.; Rui, H.; Shi, D.; Peng, X.; Ji, L.; Song, X. Lithium recovery from effluent of spent lithium battery recycling process using solvent extraction. *J. Hazard. Mater.* **2020**, *398*, 122840. [CrossRef] [PubMed]

3. Thompson, D.L.; Hartley, J.M.; Lambert, S.M.; Shiref, M.; Harper, G.D.; Kendrick, E.; Anderson, P.; Ryder, K.S.; Gaines, L.; Abbott, A.P. The importance of design in lithium ion battery recycling–a critical review. *Green Chem.* **2020**, *22*, 7585–7603. [CrossRef]

4. Mohr, M.; Peters, J.F.; Baumann, M.; Weil, M. Toward a cell-chemistry specific life cycle assessment of lithium-ion battery recycling processes. *J. Ind. Ecol.* **2020**, *24*, 1310–1322. [CrossRef]

5. Bai, Y.; Muralidharan, N.; Sun, Y.-K.; Passerini, S.; Whittingham, M.S.; Belharouak, I. Energy and environmental aspects in recycling lithium-ion batteries: Concept of Battery Identity Global Passport. *Mater. Today* **2020**, *41*, 304–315. [CrossRef]

6. Sloop, S.; Crandon, L.; Allen, M.; Koetje, K.; Reed, L.; Gaines, L.; Sirisaksoontorn, W.; Lerner, M. A direct recycling case study from a lithium-ion battery recall. *Sustain. Mater. Technol.* **2020**, *25*, e00152. [CrossRef]

7. Buss, S.R. Introduction to inverse kinematics with jacobian transpose, pseudoinverse and damped least squares methods. *IEEE J. Robot. Autom.* **2004**, *17*, 16.

8. Dang, H.; Wang, B.; Chang, Z.; Wu, X.; Feng, J.; Zhou, H.; Li, W.; Sun, C. Recycled Lithium from Simulated Pyrometallurgical Slag by Chlorination Roasting. *ACS Sustain. Chem. Eng.* **2018**, *6*, 13160–13167. [CrossRef]

9. Vieceli, N.; Nogueira, C.A.; Guimarães, C.; Pereira, M.F.C.; Durão, F.O.; Margarido, F. Hydrometallurgical recycling of lithium-ion batteries by reductive leaching with sodium metabisulphite. *Waste Manag.* **2018**, *71*, 350–361. [CrossRef]

10. Wang, H.; Friedrich, B. Development of a Highly Efficient Hydrometallurgical Recycling Process for Automotive Li–Ion Batteries. *J. Sustain. Metall.* **2015**, *1*, 168–178. [CrossRef]

11. Gaines, L. The future of automotive lithium-ion battery recycling: Charting a sustainable course. *Sustain. Mater. Technol.* **2014**, *1–2*, 2–7. [CrossRef]

12. Tanong, K.; Blais, J.-F.; Mercier, G. Metal recycling technologies for battery waste. *Recent Pat. Eng.* **2014**, *8*, 13–23. [CrossRef]

13. Heelan, J.; Gratz, E.; Zheng, Z.; Wang, Q.; Chen, M.; Apelian, D.; Wang, Y. Current and prospective Li-ion battery recycling and recovery processes. *Jom* **2016**, *68*, 2632–2638. [CrossRef]

14. Wegener, K.; Chen, W.H.; Dietrich, F.; Dröder, K.; Kara, S. Robot Assisted Disassembly for the Recycling of Electric Vehicle Batteries. *Procedia CIRP* **2015**, *29*, 716–721. [CrossRef]

15. Available online: https://www.ornl.gov/news/automated-disassembly-line-aims-make-battery-recycling-safer-faster (accessed on 23 June 2022).

16. Choux, M.; Marti Bigorra, E.; Tyapin, I. Task Planner for Robotic Disassembly of Electric Vehicle Battery Pack. *Metals* **2021**, *11*, 387. [CrossRef]

17. Gerbers, R.; Wegener, K.; Dietrich, F.; Dröder, K. Safe, Flexible and Productive Human-Robot-Collaboration for Disassembly of Lithium-Ion Batteries. In *Recycling of Lithium-Ion Batteries. Sustainable Production, Life Cycle Engineering and Management*; Kwade, A., Diekmann, J., Eds.; Springer: Berlin/Heidelberg, Germany, 2018. [CrossRef]

18. Doughty, D.H.; Roth, E.P. A general discussion of Li ion battery safety. *Electrochem. Soc. Interface* **2012**, *21*, 37–44.

19. Kong, L.; Li, C.; Jiang, J.; Pecht, M.G. Li-ion battery fire hazards and safety strategies. *Energies* **2018**, *11*, 2191. [CrossRef]

20. Nakamura, Y.; Hanafusa, H. Inverse kinematic solutions with singularity robustness for robot manipulator control. *J. Dyn. Sys. Meas. Control.* **1986**, *108*, 163–171. [CrossRef]

21. Spong, M.W.; Hutchinson, S.; Vidyasagar, M. *Robot Modeling and Control*; Wiley: Hoboken, NJ, USA, 2006.

22. FANUC. *FANUC Robot LR-Mate 200iD*; FANUC America Corporation: Rochester Hills, MI, USA, 2013; pp. 48309–48325. Available online: https://www.fanucamerica.com/docs/default-source/robotics-product-information-sheets/lr-mate-200id-series_187.pdf?sfvrsn=acc1fb06_4 (accessed on 10 April 2019).

23. Wampler, C.W. Manipulator inverse kinematic solutions based on vector formulations and damped least-squares methods. *IEEE Trans. Syst. Man Cybern.* **1986**, *16*, 93–101. [CrossRef]

24. Kay, I.; Esmaeeli, R.; Hashemi, S.R.; Mahajan, A.; Farhad, S. Recycling Li-ion Batteries: Robot Disassembly of Electric Vehicle Battery Systems, IMECE2019-11949. In Proceedings of the ASME 2019, International Mechanical Engineering Congress and Exposition (IMECE), Salt Lake City, UT, USA, 11–14 November 2019.

25. Al-Shammari, H.; Farhad, S. Performance of Cathodes Fabricated from Mixture of Active Materials Obtained from Recycled Lithium-Ion Batteries. *Energies* **2022**, *15*, 410. [CrossRef]

26. Al-Shammari, H.; Farhad, S. Heavy Liquids for Rapid Separation of Cathode and Anode Active Materials from Recycled Lithium-Ion Batteries. *Resour. Conserv. Recycl.* **2021**, *174*, 105749. [CrossRef]

27. Farhad, S.; Gupta, R.K.; Yasin, G.; Nguyen, T.A. *Nanotechnology for Battery Recycling, Remanufacturing, and Reusing*; Elsevier: Amsterdam, The Netherlands, 2022; ISBN 13978-0323911344.

Article

Modal Analysis of a Lithium-Ion Battery for Electric Vehicles

Nicholas Gordon Garafolo *, Siamak Farhad *, Manindra Varma Koricherla, Shihao Wen and Roja Esmaeeli

Department of Mechanical Engineering, University of Akron, Akron, OH 44325, USA;
mk184@uakron.edu (M.V.K.); sw118@uakron.edu (S.W.); re25@uakron.edu (R.E.)
* Correspondence: ngarafo@uakron.edu (N.G.G.); sfarhad@uakron.edu (S.F.)

Abstract: The battery pack in electric vehicles is subjected to road-induced vibration and this vibration is one of the potential causes of battery pack failure, especially once the road-induced frequency is close to the natural frequency of the battery when resonance occurs in the cells. If resonance occurs, it may cause notable structural damage and deformation of cells in the battery pack. In this study, the natural frequencies and mode shapes of a commercial pouch lithium-ion battery (LIB) are investigated experimentally using a laser scanning vibrometer, and the effects of the battery supporting methods in the battery pack are presented. For this purpose, a test setup to hold the LIB on the shaker is designed. A numerical analysis using COMSOL Multiphysics software is performed to confirm that the natural frequency of the designed test setup is much higher than that of the battery cell. The experimental results show that the first natural frequency in the two-side supported and three-side supported battery is about 310 Hz and 470 Hz, respectively. Although these frequencies are more than the road-induced vibration frequencies, it is recommended that the pouch LIBs are supported from three sides in battery packs. The voltage of the LIB is also monitored during all experiments. It is observed that the battery voltage is not affected by applying mechanical vibration to the battery.

Keywords: lithium-ion battery; modal analysis; electric vehicles; vibration; experiments

Citation: Garafolo, N.G.; Farhad, S.; Koricherla, M.V.; Wen, S.; Esmaeeli, R. Modal Analysis of a Lithium-Ion Battery for Electric Vehicles. *Energies* **2022**, *15*, 4841. https://doi.org/10.3390/en15134841

Academic Editor: Alessandro Del Nevo

Received: 9 June 2022
Accepted: 22 June 2022
Published: 1 July 2022

Publisher's Note: MDPI stays neutral with regard to jurisdictional claims in published maps and institutional affiliations.

1. Introduction

Relatively high energy and power densities, long cycle life, lightweight design, and the absence of memory effect have facilitated lithium-ion batteries (LIBs) have expanded in use as a sustainable product for supplying electrical energy [1–3]. The use of LIBs in electric vehicles (EVs) and hybrid electric vehicles (HEVs) is also becoming more attractive to automobile manufacturers. LIBs are preferred over other battery technologies for EVs and HEVs due to their higher energy and power densities. At present, substantial research and development efforts are being focused on increasing the range of applications of LIBs in EVs and HEVs.

The study of vibration and impact loads of LIBs and their casing is of great importance to avoid their malfunctioning in EV and HEV operations, which has led to commitments from researchers to study the behavior of LIBs under such conditions. Concerning the safety of the vehicle and, consequently, its passengers, the probability of short circuit or thermal runaway, or explosion in its worst-case, shall be studied thoroughly. Consequently, a lot of researchers have done studies on quantifying the mechanical robustness of the cell and investigating, for instance, the effect of nail penetration [4], the mechanical crushing of the cell [5–7], the durability of the cell under impact [5–8], mechanical shock [5] and fatigue conditions of the battery system under extreme temperature and pressure conditions [9,10]. These are done to satisfy the requirements of the whole vehicle crash homologation standards which are reviewed specifically for the use of LIBs in transportation by Refs. [11–13]. In addition, some researchers examined battery shell casing properties for cylindrical LIBs for the EV crash [5], and performed axial and lateral compression tests, three points bend tests, and hydraulic tests [14]. Furthermore, the battery brackets are studied by a means of single-axis acceleration test approach [15]. Therefore, conducting investigations on the amplitude and frequency of the vibration to which batteries are exposed is vital to understanding how the mechanical vibration affects electronic and electrical components and how to avoid or hinder the loss of electrical continuity and the housing structural failure, which is a common well-known cause of failure [15–19]. It has been demonstrated that structural failure with sustained and excessive motion inevitably happens if a system vibration happens at the same frequency as its natural frequency [16]. Some studies are dedicated to studying individual single-cell responses to vibration while some referred to considering the whole battery pack. Choi et al. [20], investigated the natural frequencies of a used but serviceable 10 Ah LIB pouch cell during impact hammer and shaker excitation tests. The results of this study showed that the first and second mode shape frequencies of the battery to the impact hammer testing are roughly 267 Hz and 474 Hz, respectively. For the frequency response of the battery for the white noise wave to the shaker, with two-edge clamping condition along the out-of-plane excitation, the first and second mode shapes are around 87 Hz and 478 Hz, respectively. Similarly, Hoopen and James [17] studied impulse excitation provided by an impact hammer to quantify the natural frequencies and mode shapes of a commercially available 25 Ah Nickel Manganese Cobalt Oxide (NMC) laminate pouch cell. However, compared to Choi et al.'s study, they covered the calculation of the values of cell damping and stiffness and reported that the lowest natural frequency of the cell is about 200 Hz [20].

There are three methods of excitations commonly employed to measure the modal response of a structure: impulse excitation, dynamic excitation, and operational excitation [21]. The impulse excitation is implemented using a hammer applying the impact testing. This method is commonly employed when the testing of specific points is aimed. However, the destruction of the structure during the test should be considered. In operational excitation, the structure is subjected to the real-life application and it is done after the structure is subjected to impulse and dynamic excitation and acts as a final verification test [21]. However, the low accuracy of the operational excitement to measure the natural frequency of a battery system in EVs or HEVs is its main drawback. Dynamic excitation is often employed for structures for which performing the other two methods is not applicable, since they are destructive testing. The way that dynamic excitation is employed is to use an electromagnetic shaker or hydraulic shaker to apply a force or frequency input that is known to the structure. The flexibility that dynamic excitation offers is that researchers can apply different values of inputs conveniently, such as specific force and specific frequency [21].

The study of the vehicle vibration inputs to the batteries suggests testing EV and HEV battery vibrational characteristics over the range of 0 Hz to 150 Hz frequency [17]. In addition, standards and regulations recommend testing in the range of 7–200 Hz, 7–50 Hz, and 10–190 Hz for the regulation UN38.3 [22], regulation ECE 100 [23] and the SAE J2380 standard [24], respectively. Consequently, when the natural frequency of the battery and the support are beyond 200 Hz the battery may not experience the failure due to vibration. This paper suggests an experimental method to quantify a commercial LIB's natural frequencies and mode shapes. The laser scanning vibrometer is used for modal analysis with frequency response functions (FRF). The design of the battery holder to test the battery in supporting configurations of two-side and three-side clamping is presented. The effect of the battery on the LIB voltage fluctuation is also studied.

2. Formulations

COMSOL Multiphysics software is employed to calculate the natural frequency of the base-plate design and to ensure that it does not coincide with the natural frequency of the battery and the excitation frequency. The un-damped free vibration equation for the system is as follows:

$$M\ddot{X} + SX = 0 \tag{1}$$

where, M is the mass matrix, $\ddot{X}$ is the mass acceleration vectors, S is the stiffness matrix, and X is the displacement vectors of the modes. In a natural mode of vibration, the displacement of each mode is calculated by:

$$X_i = X_{i,m} \times \sin(\omega t + \varnothing_i) \tag{2}$$

where, ω and $\varnothing_i$ are the angular frequency and phase angle of the ith mode. X_i is the matrix of the displacement of the modes, and $X_{i,m}$ is the vector of maximum values. If the displacement field of the given structure is harmonic, the Eigen frequency can be derived. Dictating equations in the study are in terms of the excitation load.

$$-\rho\omega^2 u = \nabla \times FS + F_v \tag{3}$$

where, ρ is the density of the material, ω is the angular frequency of the excitation load, and u is the harmonic response from the structure. The Eigenvalues λ and the Eigen frequencies f are calculated using Equation (4):

$$f = \left| \frac{Im\,(i\omega)}{2\pi} \right| \tag{4}$$

where, the eigenvalue is $\lambda = -i\omega$.

3. Experimental

The design of the test apparatus is done by SolidWorks software, where the LIB is held on the shaker and the baseplate is designed in a way to accommodate the geometry of the battery. To replicate the real-life mounting conditions of the battery, the baseplate aims to provide rigid support to the battery and hold the battery firmly. The baseplate is designed such that the battery fits in easily and the fixture including the battery does not exceed the weight-bearing limit of the shaker. The material used for the baseplate is 6061 Aluminum and its geometry is presented in Figure 1. The baseplate is mounted on the shaker using M6 screws to the center of the shaker and is torqued down with 45 lb/in. Then, the battery is fixed on that plate with clips. To perform sinusoidal frequency sweeps, a 110 lb MB RED dynamic shaker is used. A signal generator is used to create input variables. The shaker specifications are listed in Table 1. Due to restrictions of weight that the dynamic shaker aperture load is 12 lb and the maximum weight of the apparatus that the aperture arm of the shaker can handle is 11 lb, the weight of the fixture and apparatus including the battery is determined to be 10 lb. At the test of the structure mounted aperture arm, the dynamic shaker delivers low noise motion. The casing material used all around the flexures to hold the internal components is stainless steel. Using a set of ultra-flexible multi-strand wire, coil currents are conducted to the coil from which the shaker receives the signal and responds accordingly. The cooling system is provided with a constant field and eliminates the need for a power source. The reason for using a cooling system is to reduce the resistive losses of the electromagnet from coil overheating and abate the breakdown of the coil insulation [25]. The baseplate is then installed onto the aperture arm of the shaker. The completed experimental setup with the battery is presented in Figure 2.

A commercially available aftermarket pouch LIB consisting of 23 individual cells that are sandwiched together, with the specifications listed in Table 2, is used. We choose the pouch type for the LIB because it is getting more attraction for automobile applications or heavy-duty vehicles.

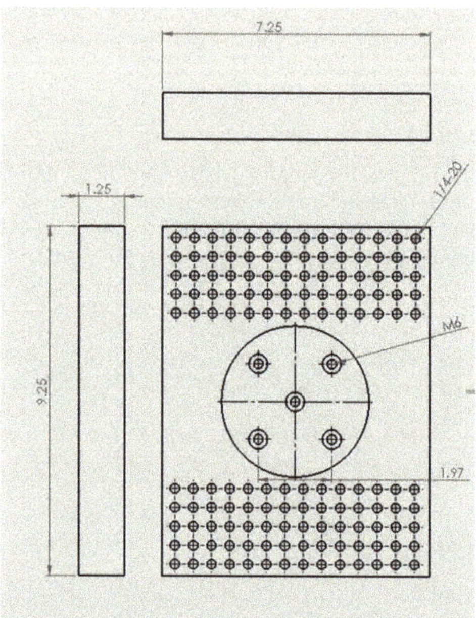

Figure 1. Drawing of the plate in SolidWorks Software.

Table 1. Specification of the energizer red shaker shown in Figure 2.

Specifications	Value
Force output-cooling (convection)	55 pounds (peak)
Stroke	1.5 inch (peak-peak)
Distance between stops	1.5 inch
Armature axial stiffness	65 pounds/inch
Armature weight	2.6 pounds
The frequency range of shaker	DC to 5000 Hz
Driver coil current	22 Amperes Maximum
Driver coil DC resistance	41 Ohms (@ amplifier connector on shaker)
Shaker attachments	Floor mount trunnion base
Dimensions	13.1 in high (to top of mounting table) 14 in × 10.75 in footprint
Weight	85 lbs.

Table 2. The specification of the aftermarket LIB used in this study.

Item	Value
The thickness of copper current collector	10 μm
The thickness of aluminum current collector	20 μm
Thickness of anode	57 μm
Thickness of cathode	65 μm
Thickness of separator	48 μm
Anode material	Graphite
Cathode material	Lithium iron phosphate, $LiFePO_4$ (LFP)

The velocity of the battery is directly measured with a Polytec PSV 400 laser scanner and the velocity data is converted to FRF calculations using integrated laser vibrometer software. For conducting calculations of the frequency response function, there is a built-in accelerometer that is attached to the surface of the dynamic shaker. The experimental setup block diagram is shown in Figure 3.

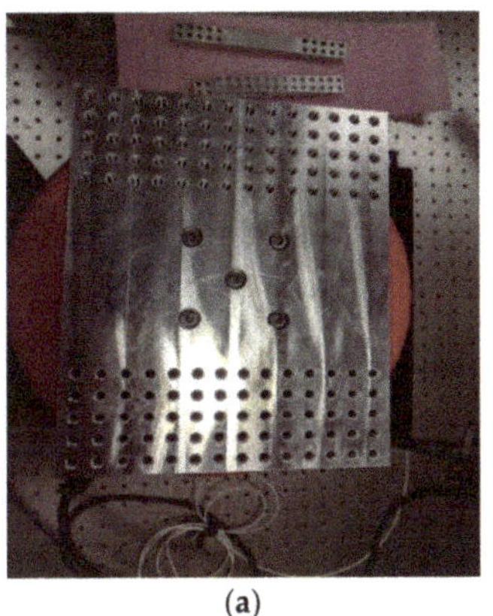
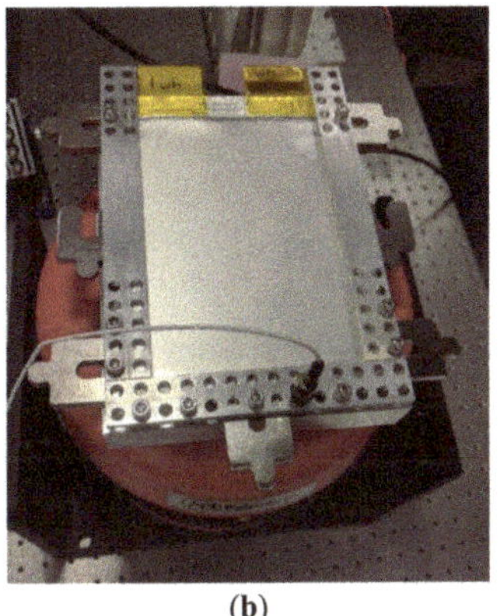

(a)   (b)

Figure 2. (**a**) Baseplate installed onto the aperture arm of the shaker, (**b**) Complete experimental setup with the pouch LIB.

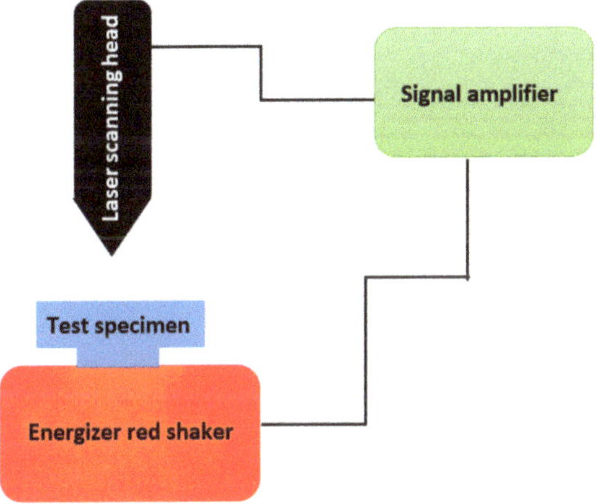

Figure 3. Block diagram of the experimental setup.

3.1. Boundary Conditions

To set up the system to represent real conditions of a battery in EVs/HEVs, two different boundary conditions are examined. Boundary condition 1 is the rigidly clamped mounting technique that fixes the three sides of the battery as shown in Figure 4a. Boundary condition 2 is the partially clamped mounting technique which can be seen in Figure 4b. This boundary condition leaves the output of the battery and the opposite side open and clamped rigidly lengthwise.

3.2. Grid Convergence

A study on grid convergence is necessary since grid densities of a scanning laser vibrometer should not have any effect on the experiments. Three different grid sizes from widely-placed to closely-placed grid points are shown in Table 3. The results from the grid convergence study in Figure 5 show that the mode shapes resulting from boundary condition 1 are consistent with the increase of scanning laser vibrometer grid densities. The scanning time of each grid size, as shown in Table 3, takes more time with an increase of grid densities. Finally, based on the results from the mode shapes and duration of the scan, evenly placed grid size mesh 2 with 425 points is chosen for the rest of the study.

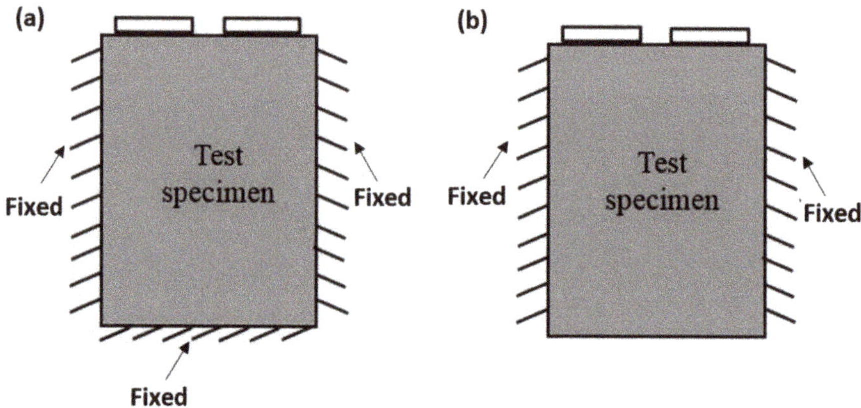

Figure 4. Battery Boundary conditions (**a**) Boundary condition 1, which is the three-side clamped, and (**b**) Boundary condition 2, which is the two-side clamped.

Table 3. Mesh representation for the mesh size independency analysis.

Mesh Number	Mesh Points	Mesh Representation	Scan Duration
Mesh 1	117		35 min
Mesh 2	425		1 h 25 min
Mesh 3	1247		6 h 20 min

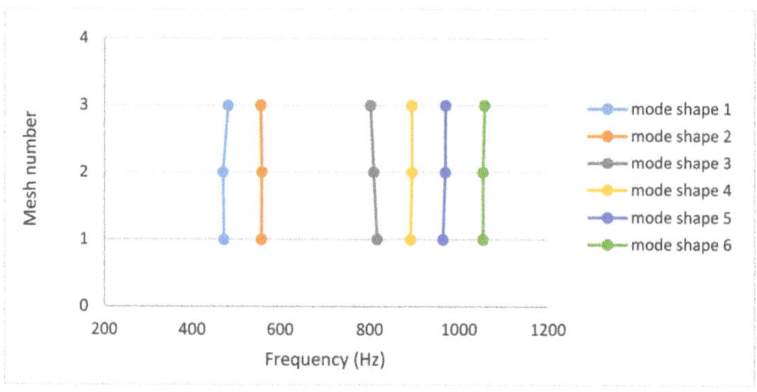

Figure 5. Each mode shape frequency for the three mesh sizes.

4. Results and Discussion

4.1. Baseplate

The action to take into consideration is to ensure that the natural frequencies of the baseplate and the battery do not happen at an equal frequency. Otherwise, this might lead to misinterpretation of the results because of the resonance that the whole apparatus will face. To verify this, an Eigen frequency study with the use of COMSOL Multiphysics software is done on the baseplate. Figure 6 shows the first mode shape of the baseplate. By gradually increasing the frequency input to the structure, there will be other natural frequencies shown by it, as presented in Table 4.

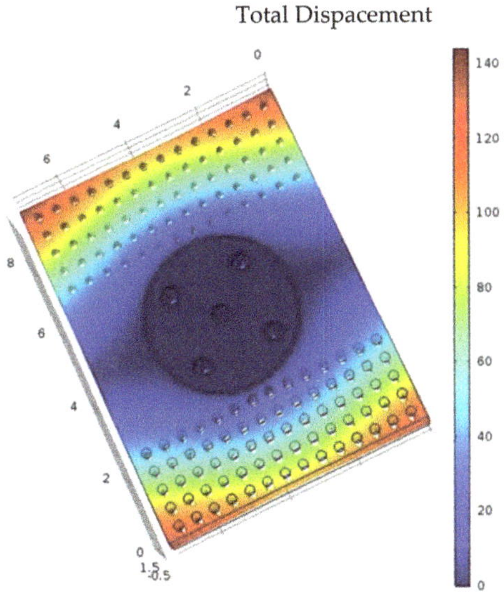

Figure 6. The first mode shape of the baseplate happens at 2716.3 Hz.

Table 4. The first six natural frequencies of the baseplate.

Frequency No.	Natural Frequency
1	2716.3 Hz
2	2999.0 Hz
3	3013.8 Hz
4	3118.5 Hz
5	4247.1 Hz
6	5398.7 Hz

The first natural frequency of the baseplate, which is the frequency in which the first bending of the structure occurs, happens at 2716.3 Hz, while the first natural frequency of the LIB is reported to be lower than 200 Hz [17,20]. Consequently, there is no match between the natural frequency of the baseplate and the battery, which means that the whole apparatus will not resonate at the same natural frequency.

4.2. Battery Mode Shapes

In Figures 7 and 8, the magnitude of mode shapes for boundary conditions 1 and 2 are shown. A range of 470–1060 Hz for the first six natural frequencies for boundary condition 1 is seen, while for boundary condition 2 the range of 310–995 Hz for the first six natural frequencies is observed. However, the vibration induced by the road does not exceed 200 Hz, which is far below the ranges discussed above. Consequently, it is unlikely that the natural frequencies coincide with road-induced vibrations. Thus, both stacking techniques can be implemented.

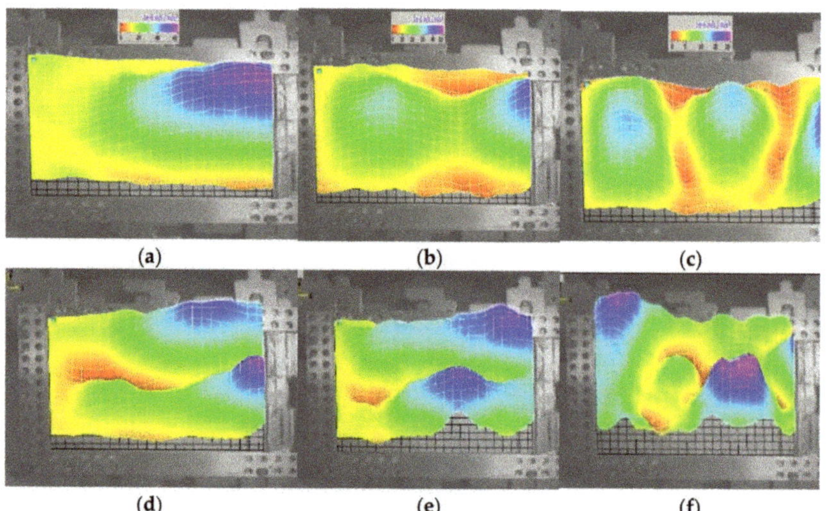

Figure 7. Pouch LIB mode shapes for boundary condition 1 shown in Figure 4a. (**a**) mode shape 1–first bending, (**b**) mode shape 2–first torsion, (**c**) mode shape 3–second torsion, (**d**) mode shape 4–second bending, (**e**) mode shape 5–third bending, (**f**) mode shape 6–third torsion.

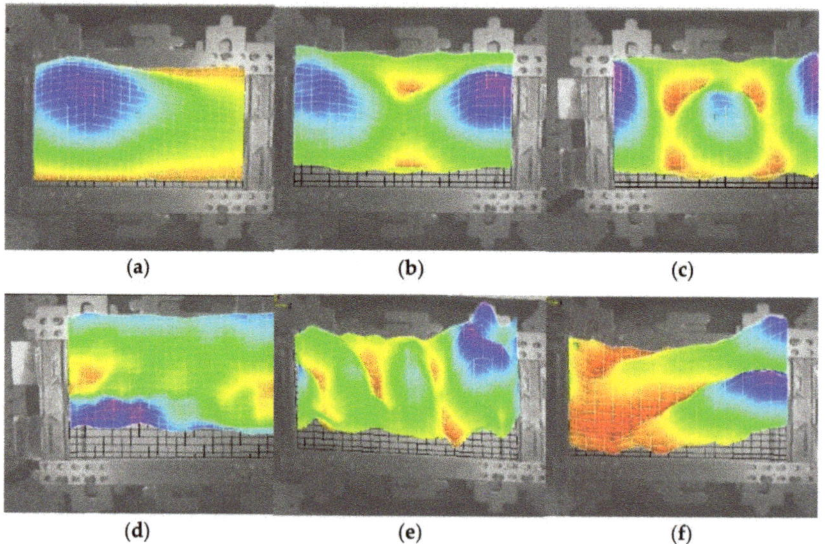

Figure 8. LIB mode shapes in boundary condition 2 shown in Figure 4b. (**a**) mode shape 1–first bending, (**b**) mode shape 2–first torsion, (**c**) mode shape 3–second torsion, (**d**) mode shape 4–second bending, (**e**) mode shape 5–third bending, (**f**) mode shape 6–third torsion.

The change in battery voltage is monitored during the vibration test. By measurement of the battery voltage in different time intervals, it is revealed that the battery voltage remains at 3.287 V with no significant fluctuation. Finally, after performing a complete test for both boundary conditions, no battery failure is observed.

4.3. Repeatability Analysis

To ensure the reliability of the frequency results and the mode shapes, a repeatability study with a sample size of 11 is done. A break is given to the test apparatus after each test. To cover all mode shapes a wide frequency range of 200–2000 Hz is chosen. The results of repeatability test runs are presented in Figure 9 for boundary conditions 1 and 2, which show that there is no significant deviation of the observed frequencies.

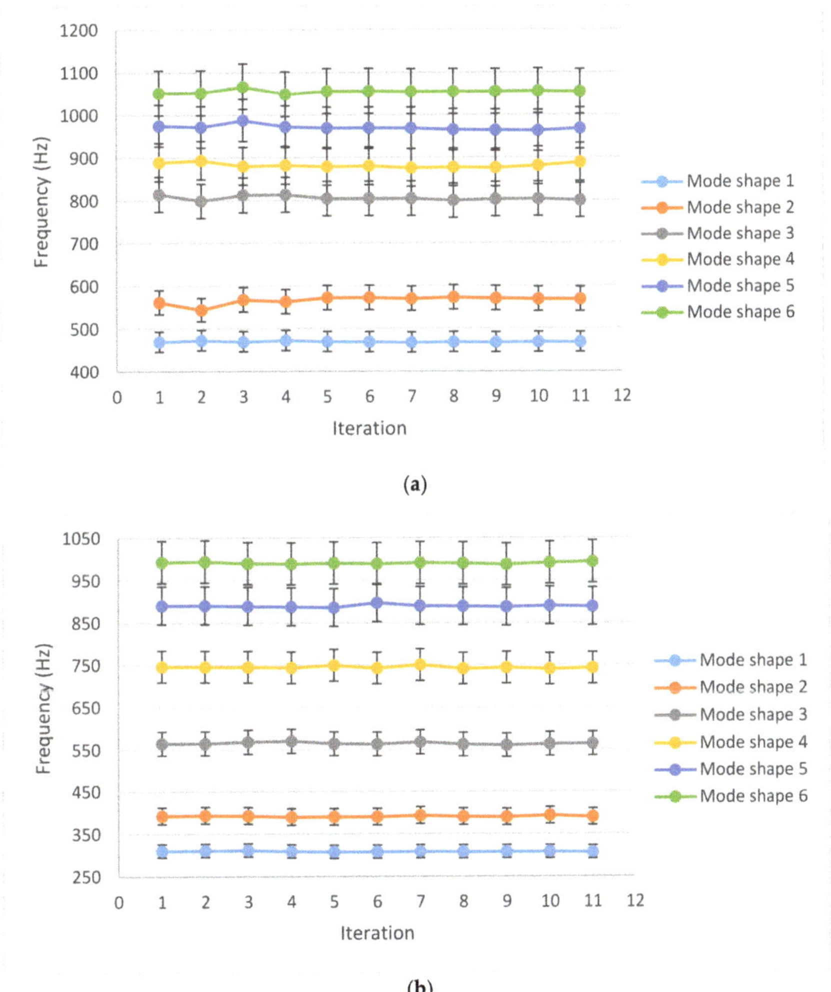

Figure 9. Mode shapes frequency for 11 repetitions; (**a**) boundary condition 1, (**b**) boundary condition 2.

5. Conclusions

The battery installed in EVs and HEVs is subjected to continuous vibration caused by road conditions. If the frequency of the road-induced vibration matches the natural frequency of the battery pack, there may be a failure in the battery due to the created resonance. In this paper we studied the mode shapes and natural frequencies of a commercially available pouch LIB, designed for EVs and HEVs. The LIB was excited by a dynamic shaker and the study was done using a laser scanning vibrometer. Two battery boundary conditions of the fixed-fixed-fixed-free (boundary condition 1) and the fixed-free-fixed-free (boundary condition 2) were examined. These two boundary conditions simulated the battery installation conditions in EVs/HEVs' battery packs. The repeatability tests and grid convergence studies were performed to ensure the accuracy and reliability of test results. The results revealed that the frequencies of the first six mode shapes of the battery fall within the range of 460–1070 Hz and 310–1000 Hz for boundary conditions 1 and 2, respectively. Although a change in boundary condition will alter the natural frequency of the battery, in either of the boundary conditions the natural frequency of the battery does not coincide with the road-induced vibration, which is lower than 200 Hz. The battery voltage was also monitored during the vibration tests. It was revealed that applying vibration with frequencies lower than the resonance frequencies will not affect the battery voltage. The results presented in this paper is for a pouch LIB with liquid electrolyte and the graphite anode and LFP cathode active materials. A separate study may be required to prove that the results of this paper can be generalized for LIBs with different designs and different electrode active materials and electrolytes.

Author Contributions: Data curation, M.V.K.; Formal analysis, N.G.G., S.F., M.V.K., S.W. and R.E.; Funding acquisition, N.G.G. and S.F.; Investigation, N.G.G., M.V.K. and R.E.; Methodology, N.G.G. and S.F.; Software, S.F.; Supervision, N.G.G. and S.F.; Writing—original draft, M.V.K., S.W. and R.E.; Writing—review & editing, N.G.G. and S.F. All authors have read and agreed to the published version of the manuscript.

Funding: This research received no external funding.

Institutional Review Board Statement: Not applicable.

Informed Consent Statement: Not applicable.

Data Availability Statement: Not applicable.

Acknowledgments: The financial support of the University of Akron is highly appreciated.

Conflicts of Interest: The authors declare no conflict of interest.

References

1. Foreman, E.; Zakri, W.; Sanatimoghaddam, M.H.; Modjtahedi, A.; Pathak, S.; Kashkooli, A.G.; Garafolo, N.G.; Farhad, S. A review of inactive materials and components of flexible lithium-ion batteries. *Adv. Sustain. Syst.* **2017**, *1*, 1700061. [CrossRef]
2. Iordache, A.; Bresser, D.; Solan, S.; Retegan, M.; Bardet, M.; Skrzypski, J.; Picard, L.; Dubois, L.; Gutel, T. From an enhanced understanding to commercially viable electrodes: The case of PTCLi$_4$ as sustainable organic lithium-ion anode material. *Adv. Sustain. Syst.* **2017**, *1*, 1600032. [CrossRef]
3. Sasi, R.; Chandrasekhar, B.; Kalaiselvi, N.; Devaki, S.J. Green solid ionic liquid crystalline electrolyte membranes with anisotropic channels for efficient li-ion batteries. *Adv. Sustain. Syst.* **2017**, *1*, 1600031. [CrossRef]
4. Feng, X.; Sun, J.; Ouyang, M.; Wang, F.; He, X.; Lu, L.; Peng, H. Characterization of penetration induced thermal runaway propagation process within a large format lithium ion battery module. *J. Power Sources* **2014**, *275*, 261–273. [CrossRef]
5. Avdeev, I.; Gilaki, M. Structural analysis and experimental characterization of cylindrical lithium-ion battery cells subject to lateral impact. *J. Power Sources* **2014**, *271*, 382–391. [CrossRef]
6. Greve, L.; Fehrenbach, C. Mechanical testing and macro-mechanical finite element simulation of the deformation, fracture, and short circuit initiation of cylindrical Lithium-ion battery cells. *J. Power Sources* **2012**, *214*, 377–385. [CrossRef]
7. Sahraei, E.; Meier, J.; Wierzbicki, T. Characterizing and modeling mechanical properties and onset of short circuit for three types of lithium-ion pouch cells. *J. Power Sources* **2014**, *247*, 503–516. [CrossRef]
8. Shi, F.; Yu, H.; Chen, X.; Cui, T.; Zhao, H.; Shi, X. Mechanical performance study of lithium-ion battery module under dynamic impact test. In *Proceedings of the 19th Asia Pacific Automotive Engineering Conference & SAE-China Congress 2017: Selected Papers*; Springer: Singapore, 2019; pp. 1–11.
9. Spinner, N.S.; Field, C.R.; Hammond, M.H.; Williams, B.A.; Myers, K.M.; Lubrano, A.L.; Rose-Pehrsson, S.L.; Tuttle, S.G. Physical and chemical analysis of lithium-ion battery cell-to-cell failure events inside custom fire chamber. *J. Power Sources* **2015**, *279*, 713–721. [CrossRef]
10. Liu, X.; Stoliarov, S.I.; Denlinger, M.; Masias, A.; Snyder, K. Comprehensive calorimetry of the thermally-induced failure of a lithium ion battery. *J. Power Sources* **2015**, *280*, 516–525. [CrossRef]
11. Huo, H.; Xing, Y.; Pecht, M.; Züger, B.J.; Khare, N.; Vezzini, A. Safety requirements for transportation of lithium batteries. *Energies* **2017**, *10*, 793. [CrossRef]
12. Farrington, M.D. Safety of lithium batteries in transportation. *J. Power Sources* **2001**, *96*, 260–265. [CrossRef]

13. Ruiz, V.; Pfrang, A.; Kriston, A.; Omar, N.; van den Bossche, P.; Boon-Brett, L. A review of international abuse testing standards and regulations for lithium ion batteries in electric and hybrid electric vehicles. *Renew. Sustain. Energy Rev.* **2018**, *81*, 1427–1452. [CrossRef]
14. Zhang, X.; Wierzbicki, T. Characterization of plasticity and fracture of shell casing of lithium-ion cylindrical battery. *J. Power Sources* **2015**, *280*, 47–56. [CrossRef]
15. Choi, Y.; Jung, D.; Ham, K.; Bae, S. A study on the accelerated vibration endurance tests for battery fixing bracket in electrically driven vehicles. *Procedia Eng.* **2011**, *10*, 851–856. [CrossRef]
16. Moon, S.-I.; Cho, I.-J.; Yoon, D. Fatigue life evaluation of mechanical components using vibration fatigue analysis technique. *J. Mech. Sci. Technol.* **2011**, *25*, 631–637. [CrossRef]
17. Hooper, J.M.; Marco, J. Experimental modal analysis of lithium-ion pouch cells. *J. Power Sources* **2015**, *285*, 247–259. [CrossRef]
18. Hooper, J.M.; Marco, J. Characterising the in-vehicle vibration inputs to the high voltage battery of an electric vehicle. *J. Power Sources* **2014**, *245*, 510–519. [CrossRef]
19. Hooper, J.M.; Marco, J.; Chouchelamane, G.H.; Chevalier, J.S.; Williams, D. Multi-axis vibration durability testing of lithium-ion 18650 NCA cylindrical cells. *J. Energy Storage* **2018**, *15*, 103–123. [CrossRef]
20. Choi, H.Y.; Lee, I.; Lee, J.S.; Kim, Y.M.; Kim, H. A study on mechanical characteristics of lithium-polymer pouch cell battery for electric vehicle. In Proceedings of the 23rd International Technical Conference on the Enhanced Safety of Vehicles (ESV) National Highway Traffic Safety Administration, Seoul, Korea, 27–30 May 2013.
21. Ewins, D.J. Basics and state-of-the-art of modal testing. *Sadhana* **2000**, *25*, 207–220. [CrossRef]
22. Available online: www.mtixtl.com/UN38.3/UN_Test_Manual_Lithium_Battery_Requirements.PDF (accessed on 23 June 2022).
23. Available online: https://unece.org/fileadmin/DAM/trans/main/wp29/wp29regs/2013/R100r2e.pdf (accessed on 23 June 2022).
24. Available online: www.sae.org/standards/content/j2380_201312/ (accessed on 23 June 2022).
25. User Manual of Red Energizer, MB Dynamics. Available online: https://www.mbdynamics.com/wp-content/uploads/2018/10/MB-BSR-ENRGZR-RED-1018-FINAL.pdf (accessed on 23 June 2022).

Article

Design of a Non-Linear Observer for SOC of Lithium-Ion Battery Based on Neural Network

Ning Chen, Xu Zhao, Jiayao Chen *, Xiaodong Xu *, Peng Zhang and Weihua Gui

School of Automation, Central South University, Changsha 410083, China; ningchen@csu.edu.cn (N.C.); zhaoxu633@163.com (X.Z.); zzppawj@163.com (P.Z.); gwh@csu.edu.cn (W.G.)
* Correspondence: jychen@csu.edu.cn (J.C.); xiaodongxu@csu.edu.cn (X.X.)

Abstract: This paper presents a method for use in estimating the state of charge (SOC) of lithium-ion batteries which is based on an electrochemical impedance equivalent circuit model with a controlled source. Considering that the open-circuit voltage of a battery varies with the SOC, an equivalent circuit model with a controlled source is proposed which the voltage source and current source interact with each other. On this basis, the radial basis function (RBF) neural network is adopted to estimate the uncertainty in the battery model online, and a non-linear observer based on the radial basis function of the RBF neural network is designed to estimate the *SOC* of batteries. It is proved that the *SOC* estimation error is ultimately bounded by Lyapunov stability analysis, and the error bound can be arbitrarily small. The high accuracy and validity of the non-linear observer based on the RBF neural network in *SOC* estimation are verified with experimental simulation results. The *SOC* estimation results of the extended Kalman filter (EKF) are compared with the proposed method. It improves convergence speed and accuracy.

Keywords: lithium-ion batteries; state of charge; fraction order; electrochemical impedance model; non-linear observer based on RBF neural network

Citation: Chen, N.; Zhao, X.; Chen, J.; Xu, X.; Zhang, P.; Gui, W. Design of a Non-Linear Observer for SOC of Lithium-Ion Battery Based on Neural Network. *Energies* **2022**, *15*, 3835. https://doi.org/10.3390/en15103835

Academic Editor: Siamak Farhad

Received: 15 April 2022
Accepted: 18 May 2022
Published: 23 May 2022

Publisher's Note: MDPI stays neutral with regard to jurisdictional claims in published maps and institutional affiliations.

1. Introduction

With the shortage of resources and environmental pollution gradually becoming more serious, people are paying more attention to Battery Electric Vehicles (BEVs) due to their advantages of environmental protection, clean energy, low cost, high technology [1]. In order to ensure the safe and reliable operation of BEVs, the following items are essential components of BEVs: increasing the driving range, extending the battery life, reducing the battery cost [2–4], and a Battery Management System (BMS). *SOC* estimation is one of the main functions of a BMS [5]. Its value is directly related to battery life and safety. However, *SOC* cannot be measured directly, it can only be estimated by measuring variables such as current and voltage [6].

A battery model with small error and high accuracy is required, which can not only improve the accuracy of battery state estimation but also provide accurate monitoring of the important status of batteries [7]. Many methods have been proposed to estimate battery *SOC*, which can be divided into four main methods: open-circuit voltage methods, current time integration methods [8], data-driven methods [9,10], and model-driven methods. However, the first three methods have disadvantages such as large estimation errors, low reliability, and being time-consuming [11]. Therefore, the model-based methods have been studied by researchers because of their high interpretability and accuracy. The integer-order model is the most widely used in *SOC* estimation. For example, Chen et al. [12] proposed an improved Thevenin model to obtain a more accurate battery model. Considering environmental influence, Chen et al. [13] improved the battery model with the Butler–Volmer (BV) equation, which can solve the problem of large current and temperature variation.

However, compared to integer-order components, unique fractional-order components, such as CPE, can better describe the amplitude–frequency characteristics of the double layer inside the battery. For example, Liu et al. [14] proposed a fractional-order model based on the PNGV model, which considers the effects of many battery characteristics. Liu et al. [15] proposed a simplified fractional-order equivalent circuit model (FO-ECM) with high precision, which can better characterize the non-linear dynamic behaviors. However, the electrochemical process of a battery is not completely considered in this model. Electrochemical impedance spectroscopy (EIS) is widely used to establish fractional-order models, which combines both electrochemical reactions inside batteries and *SOC* estimation. Li et al. [16] proposed a *SOC* estimation method based on an equivalent circuit model in which the parameters are determined using simulated electrochemical impedance spectroscopy. Then, Hu et al. [17] established a fractional-order equivalent circuit model of lithium-ion batteries based on electrochemical impedance spectroscopy using constant-order and fractional-order characteristics such as Warburg. To improve the accuracy, Xiong et al. [18] proposed an online parameter identification method based on a fractional impedance model, which obtains the solid electrolyte interphase resistance to predict the remaining capacity. On this basis, Mawonou et al. [19] proposed a time-domain and frequency-domain fractional-order model parameter identification method based on a recursive least squares algorithm. It was proved that the fractional-order model had better performance and robustness than the classical integer-order model.

Many non-linear state estimation algorithms and adaptive filters have been studied. Typical algorithms are the Kalman filter [20,21], Luenberger observer [22], PI (proportional-integral) observer [23], H-infinity observer [24,25], sliding mode observer [26,27], particle filter [28–30], etc. In the *SOC* estimation methods based on the power battery model, the most common method is the Kalman filter, which is often used to estimate *SOC* based on linear battery models, but most battery models are non-linear. Therefore, many non-linear state estimation algorithms and adaptive filters are proposed to estimate or infer the internal state of batteries. To determine the non-linear characteristics of the battery mode, the improved Kalman filter has been further studied. Charkhgard et al. [31] proposed a method of *SOC* estimation combining a neural network (NN) and an extended Kalman filter (EKF) for *SOC* estimation. However, this method is based on a data-driven battery model, which has the disadvantage of poor interpretability. Focusing on this problem, Chen et al. [32] designed a *SOC* estimation method based on an EKF and a non-linear battery model which divides the model into a non-linear, open-circuit voltage and a second-order resistance-capacitance model. To improve the accuracy of the EKF, Zhang et al. [33] proposed a method that combines an unscented Kalman filter (UKF) and a comprehensive battery model. The system state variables are calculated recursively through a non-linear mapping process, which effectively reduces the error. However, the basic principle of the EKF algorithm is to linearize the non-linear function via first-order Taylor series expansion. This local linearization will result in large errors in the battery model when it is highly non-linear, which has limitations.

Therefore, to further improve the accuracy of *SOC* estimation, a non-linear *SOC* estimation observer was widely researched. Ouyang et al. [34] proposed an *SOC* non-linear observer based on the equivalent circuit model. The capacitance and resistance in the battery model were considered as a non-linear function of *SOC* and battery temperature, and the non-linear relationship between open-circuit voltage (*OCV*) and *SOC* was considered. However, parameter uncertainty and unknown disturbance are also key issues that affect the accuracy of *SOC* estimation most. In this case, a non-linear observer-based on a neural network was proposed. Chen et al. [35] designed an output feedback adaptive neural network controller combining adaptive backstepping technology with the approximation ability of the radial basis function neural network, which solves the problem of non-linear and non-strict systems. Zhao et al. [36] proposed an adaptive neural-network-based control scheme, in which a radial basis function neural network was adopted to compensate for the actuator gain fault. To further improve the accuracy of the *SOC* estimation, Chen et al. [37] estimated the uncertainty in the battery model online through the neural network and designed a non-linear observer based on the radial basis function neural network to estimate the *SOC* of the battery, which was proved to have faster convergence speed and higher precision.

The above methods only focus on integer-order models and take no consideration of the characteristics of fractional-order models. In this paper, a non-linear observer for *SOC* estimation with an RBF neural network, considering characteristics of lithium-ion batteries, and a fractional-order model are developed. The main contributions in this paper are (1) a fractional-order electrochemical impedance equivalent circuit model of a controlled source with the interaction of voltage source and current source is established based on electrochemical impedance spectroscopy; (2) a non-linear observer for *SOC* estimation with an RBF neural network is improved to estimate the uncertainty in a battery based on the fractional-order model. The remaining parts of this paper are organized as follows. Section 2 provides the fractional-order electrochemical impedance equivalent circuit model for lithium-ion batteries. Section 3 introduces the non-linear observer for *SOC* estimation based on a neural network. Section 4 presents the results of the simulation to verify the feasibility and superiority of the model. Finally, the conclusion is delivered in Section 5.

2. Design of Battery Model

When building a battery model, the uncertainty of the model is often large due to various reasons, and the error of the *SOC* estimation is large. In this paper, using the uncertainty of an RBF neural network estimated in a battery model online, the *SOC* of a battery could be estimated more accurately. In order to accurately estimate the *SOC* of a battery, an intuitive and comprehensive battery equivalent circuit model was adopted.

In the modeling process of lithium-ion batteries, the characterization parameters used are the voltage, the current and the temperature of the battery. These parameters are acquired by the method of time domain data acquisition. Compared with time domain signals such as voltage, current and temperature, the impedance-frequency relationship of lithium-ion battery in a specific frequency range constitutes EIS (Electrochemical impedance spectroscopy). EIS impedance spectroscopy is a non-invasive electrochemical analysis method. Its measurement results can not only analyze the performance degradation and aging state of batteries, but also provide data support for battery modeling and low-temperature heating strategies. It is related to multiple parameters, such as *SOC* estimation, internal temperature monitoring, SOH life estimation and internal failure prediction. In the EIS measurement process, the general frequency band covers 0.1Hz–10kHz, the excitation and response signals are weak and usually less than 5mV, the measurement period is long and generally within tens of seconds and minutes, the signal-to-noise ratio is below -85dB, and the time window of measurement is short be less than or equal 1 second. So EIS measurement method is too complicated, and the online measurement is difficult.

As shown in Figure 1, a typical EIS impedance spectrum is usually displayed on a negative Nyquist Diagram, with the real part of the impedance on the X-axis and the negative imaginary part of the impedance on the Y-axis. With X-axis as reference, the left side is the high frequency band. The intersection of this curve and the abscissa is the ohmic resistance of the lithium battery; the middle is the medium frequency band. The shape and parameters of this curve are formed at the interface between the lithium battery motor and the electrolyte. The electric double layer is related and is dominated by the RC parallel connected circuit, which includes the constant phase angle element (CPE); the right side is the low frequency band, and the EIS curve is a straight line with a constant slope, which is closely related to the interior of the lithium-ion active material particles. related to the solid diffusion process. It can be seen from the curve shape of EIS that lithium-ion batteries have typical fractional-order characteristics, while in the analysis process of EIS, a more accurate expression can be obtained by replacing the RC parallel component by fractional-order components. Therefore, in the ordinary second-order RC circuit model, the two capacitors are replaced by fractional-order components, and the battery model will be more accurate. The electrochemical impedance equivalent model is shown in Figure 2.

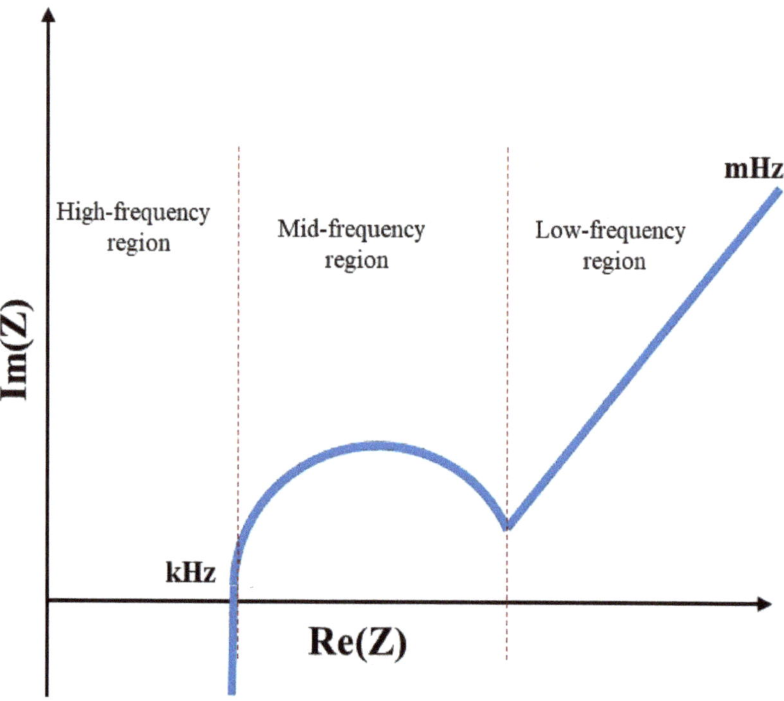

Figure 1. Typical EIS of a lithium-ion battery.

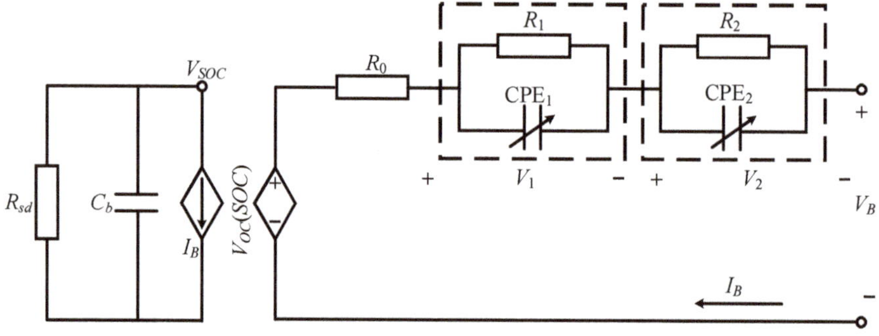

Figure 2. Electrochemical impedance circuit model of battery.

Figure 2 shows a battery electrochemical impedance equivalent circuit model with two subcircuits that interact with each other with a voltage-controlled voltage source and current-controlled current source. The left circuit in Figure 1 is used to simulate the *SOC* of a battery and the remaining operating time. The capacitance C_b is used to represent the total charge stored in a battery, and the R_{sd} is the self-discharge resistance. The capacitance C_b value is the rated capacity of the battery Q_N: $C_b = Q_N$. When the battery is charged, the charge amount on the capacitance C_b is set as Q and the charging current as I. In addition, $SOC = \int_0^t i(t)dt / Q_N = Q/Q_N = C_b \cdot V_{SOC}/Q_N = V_{SOC}$ is obtained by $Q = C_b \cdot V_{SOC}$.

In Figure 2, V_{SOC} quantitatively represents the *SOC* of batteries, and $V_{SOC} \in [0\ V, 1\ V]$ corresponds to 0–100% of the battery *SOC*. *Voc(SOC)* is the *OCV* of batteries. The nonlinear mapping of *SOC* to *OCV* of batteries $V_{OC} = f(V_{SOC}) = f(SOC)$ can also be fitted with polynomial function. V_B is a directly measurable battery end voltage. R_0, R_1 and R_2 are ohm resistors, and V_1 and V_2 are the terminal voltages of CPE_1 and CPE_2.

The mathematical model of the battery model shown in Figure 2 is established as follows:

$$\begin{cases} D^{r_1}V_1 = -\frac{1}{R_1(SOC)C_1(SOC)}V_1 + \frac{1}{C_1(SOC)}I_B \\ D^{r_2}V_2 = -\frac{1}{R_2(SOC)C_2(SOC)}V_2 + \frac{1}{C_2(SOC)}I_B \\ \dot{V}_{SOC} = -\frac{1}{R_{sd}C_b}V_{SOC} - \frac{1}{C_b}I_B \\ V_B = V_{OC} - V_1 - V_2 - R_0(SOC)I_B \end{cases} \tag{1}$$

where D_t^r is used to represent the derivative or integral of arbitrary-order r with respect to t. The Grunwald–Letnikov definition is used in this paper [38].

Considering modeling errors, sensor measurement errors and unknown interference in real processes, the above Equation (1) battery model can be rewritten as follows:

$$\begin{cases} D^r x(t) = A(x)x(t) + B(x)u(t) \\ y(t) = Cx(t) + D(x)u(t) + f(x(t)) + \varphi(x, I_B) \end{cases} \tag{2}$$

where $x(t) = [V_1(t)\ V_2(t)\ SOC(t)]^T$ is the state vector; $y(t)$ is the battery end voltage V_B which is the output of the system; $u(t)$ is the battery current I_B, which is the system input; $r = [r_1\ r_2\ 1]^T$ is the order vector of the system; $\varphi(x, I_B)$ is the uncertain item. Matrices A, B, C and D are as follows:

$$A(x_3) = \begin{bmatrix} -h_1(x_3) & 0 & 0 \\ 0 & -h_2(x_3) & 0 \\ 0 & 0 & -a \end{bmatrix}, B(x_3) = \begin{bmatrix} g_1(x_3) \\ g_2(x_3) \\ -b \end{bmatrix}, C = \begin{bmatrix} -1 & -1 & 0 \end{bmatrix}, D(x_3) = -g_3(x_3)$$

$$h_1(x_3) = \frac{1}{R_1(x_3)C_1(x_3)}, h_2(x_3) = \frac{1}{R_2(x_3)C_2(x_3)}$$

$$g_1(x_3) = \frac{1}{C_1(x_3)}, g_2(x_3) = \frac{1}{C_2(x_3)}, \tag{3}$$

$$a = \frac{1}{R_{sd}C_b}, b = \frac{1}{C_b}, g_3(x_3) = R_0(x_3)$$

$$\varphi(x, I_B) = \begin{bmatrix} 0 & 0 & \psi(x, I_B) \end{bmatrix}^T$$

The function $f(x(t))$ is widely used to represent *OCV-SOC* relationships for many batteries, which are expressed as:

$$f(x(t)) = \sum_{k=0}^{M} \mathbf{d}_k \mathbf{x}(t)^k \tag{4}$$

where $\mathbf{d}_k (k = 0, 1, \ldots M)$ is the factor of $f(\mathbf{x}(t))$. According to the relationship between *OCV-SOC* in batteries, $f(SOC)$ is a monotonic increasing function, and the easy Equation (4) is Lipschitz-continuous in the range of $0 \le SOC \le 1$, then $\|f(SOC_1) - f(SOC_2)\| \le \gamma \|SOC_1 - SOC_2\|$, where γ is the Lipschitz constant.

The coefficient of matrix $A(x)$ in Equation (3) is related to the charging and discharging time constant, whose magnitudes are generally determined in the battery system. Therefore, the time constants can be the average value of the charge and discharge time constants under different *SOCs*, which are denoted as $R_{1c}C_{1c}$, $R_{2c}C_{2c}$. In this way, Equation (2) can be rewritten as

$$\begin{cases} D^r x(t) = Ax(t) + B(x)u(t) \\ y(t) = Cx(t) + D(x)u(t) + f(x(t)) + \varphi(x, I_B) \end{cases} \tag{5}$$

where $A = \begin{bmatrix} -\frac{1}{R_{1c}C_{1c}} & 0 & 0 \\ 0 & -\frac{1}{R_{2c}C_{2c}} & 0 \\ 0 & 0 & -a \end{bmatrix} = \begin{bmatrix} -m & 0 & 0 \\ 0 & -n & 0 \\ 0 & 0 & -a \end{bmatrix}, m, n > 0.$

Remark 1. *Resistance and capacitance in an equivalent circuit can be assumed to be positive and bounded. Due to the protection of BMS, the charging and discharging current of the batteries is limited, and all states of the battery system are bounded by the non-limit current of the batteries.*

Therefore, it can be obtained that $h_1(x_3)$ and $h_2(x_3)$ in Equation (3) are positive and bounded, and $I_{B(t)}$, $x(t)$ and $g_1(x_3)$, $g_2(x_3)$, $g_3(x_3)$ are bounded.

3. Design of Non-Linear State Observer Based on RBF Neural Network

Based on improved fractional-order electrochemical impedance equivalent circuit model with a controlled source and electrochemical impedance spectroscopy in Section 2, a non-linear observer based on a neural network, where an RBF neural network is designed to approximate the uncertainty in the battery model online, is shown in this section.

The model is shown in Equation (6):

$$\begin{cases} D^r \hat{x}(t) = A\hat{x}(t) + B(\hat{x})u(t) + L(y - \hat{y}) \\ \hat{y}(t) = C\hat{x}(t) + D(\hat{x})u(t) + f(\hat{x}(t)) + \hat{\varphi}(\hat{x}, I_B) \end{cases} \tag{6}$$

In the above equation, $\hat{x}(t)$ is state estimation and $\hat{y}(t)$ is the output estimation of the actual terminal voltage. $L = [l_1, l_2, l_3]^T$ is the observer gain vector to be designed. $\hat{\varphi}(\hat{x}, I_B) = \hat{\psi}(\hat{x}, I_B)$ is the estimate value of $\varphi(x, I_B)$ in Equation (2). An RBF neural network used to estimate modeling errors and unmeasurable disturbances, as shown in Equation (7):

$$\hat{\psi}(\hat{x}, I_B) = W^T S(\hat{z}) \tag{7}$$

where the input vector is $\hat{z} = [\hat{x}^T, I_B]^T$. The weight vector is $W = [\omega_1, \omega_2, \cdots, \omega_n]^T$. The number of nodes in the neural network is N. The activation function vector is $S(\hat{z}) = [s_1(\hat{z}), \cdots, s_n(\hat{z})]^T$, where $s_i(\hat{z})(1 \leq i \leq n)$ satisfies:

$$s_i(\hat{z}) = \exp\left[\frac{-(\hat{z} - \mu_i)^T(\hat{z} - \mu_i)}{\eta_i^2}\right] \tag{8}$$

The central point vector is μ_i, and the width of the Gaussian function is η_i. The adaptive law of the weight vector is:

$$\dot{W} = \Gamma[S(\hat{z})(y - \hat{y}) - K_\omega W] \tag{9}$$

The positive definite constant matrices to be designed are $\Gamma = \Gamma^T$ and K_ω.

Remark 2. *RBF neural networks with sufficient numbers of ganglion points can approximate any continuous function. Depending on the approximation properties of RBF neural networks, the uncertainty $\psi(x, I_B)$ can be replaced by the following:*

$$\psi(x, I_B) = W^{*T} S(z) + \xi \tag{10}$$

where W^* is the true constant. The input vector is $z = [x^T, I_B]^T$. An approximate error is ξ, which is assumed to be $|\xi| \leq \xi_N$. As $\psi(\cdot)$ is bounded, and the ideal constant weight W^* is also bounded $\|W^*\| \leq W_M$, in which W_M is a normal number.

The system error dynamic equation can then be obtained and is shown in Equation (11):

$$D^r \tilde{x}(t) = \tilde{A}_{cl} \tilde{x}(t) - \left(\tilde{B} - L\tilde{D}\right) I_B(t) + LF(t) + L(\varphi(x, I_B) - \hat{\varphi}(\hat{x}, I_B)) \tag{11}$$

where the error between state estimation and state actual value is

$$\begin{aligned} &\tilde{x}(t) = x(t) - \hat{x}(t) = [\tilde{V}_1(t)\tilde{V}_2(t)S\tilde{O}C(t)]^T; \quad F(t) = f(\hat{x}(t)) - f(x(t)) \\ &\tilde{A}_{cl} = A - LC; \quad \tilde{B} = B(x) - B(\hat{x}); \quad \tilde{D} = D(x) - D(\hat{x}) \\ &\psi(x, I_B) - \hat{\psi}(\hat{x}, I_B) = W^{*T} S(x, I_B) + \xi - W^T S(\hat{x}, I_B) = -\tilde{W}^T S(\hat{x}, I_B) + \varepsilon \end{aligned} \tag{12}$$

where

$$\widetilde{W} = W - W^*, \varepsilon = W^{*T}(S(x, I_B) - S(\hat{x}, I_B)) + \xi \tag{13}$$

In Equation (13), ε is the bonded term which satisfies $|\varepsilon| \le \varepsilon_M (\varepsilon_M > 0)$. Since W^* is a constant weight vector, $\dot{\widetilde{W}} = \dot{W}$ holds.

Before giving the main theorems, the following important lemmas are given first.

Lemma 1. *A fractional differential equation, $D^{r_i} x_i(t) = g_i(t), 0 < r_i < 1$, is equivalent to the following continuous frequency distributed model.*

$$\begin{aligned} \frac{\partial z_i(\omega,t)}{\partial t} &= -\omega z_i(\omega, t) + g_i(t) \\ x_i(t) &= \int_0^\infty \mu_i(\omega) z_i(\omega, t) d\omega \\ \mu_i(\omega) &= \frac{\sin(r_i \pi)}{\pi} \omega^{-r_i} \end{aligned} \tag{14}$$

and for $r_i = 1$, $D^{r_i} x_i(t) = g_i(t)$ can be represented as

$$\begin{aligned} \frac{\partial z_i(\omega,t)}{\partial t} &= -\omega z_i(\omega, t) + g_i(t) \\ x_i(t) &= \int_0^\infty \mu_i(\omega) z_i(\omega, t) d\omega \\ \mu_i(\omega) &= \delta(\omega) \end{aligned} \tag{15}$$

where $g_i(t)$ represents the input; $x_i(t)$ denotes the output; $z_i(\omega, t)$ is the frequency distributed state variable; and $\mu_i(\omega)$ is the frequency weighting function and $\delta(\omega)$ is a unit impulse function.

Theorem 1. *If the observer gains l_1, l_2, l_3 in L given in Equation (6) to satisfy inequalities (16), the estimation error of the non-linear observer shown as Equation (6) together with the adaptive law shown in Equation (9) is uniformly bounded.*

$$\begin{cases} l_1 < m - \frac{12p_3 l_3^2}{p_1(a - \gamma l_3)} \\ l_2 < \frac{n}{5} - \frac{4p_1 l_1}{5p_1} \\ l_3 < \frac{a}{\gamma} \end{cases} \tag{16}$$

where p_1, p_2, p_3 and γ are positive constants.

Proof. According to Lemma 1, Equation (11) can be exactly converted into

$$\begin{aligned} \frac{\partial z(\omega,t)}{\partial t} &= -\omega z(\omega, t) + A_{cl} e_x(t) + Lh(e_x(t)) + (E - LF)\omega(t) \\ e_x(t) &= \int_0^\infty \mu(\omega) z(\omega, t) d\omega \end{aligned} \tag{17}$$

where

$$\begin{aligned} z(\omega, t) &= [z_1(\omega, t) \; z_2(\omega, t) \; z_3(\omega, t)]^T \\ e_x(t) &= \left[\widetilde{V}b(t) \; \widetilde{V}c(t) \; \widetilde{SOC}(t) \right]^T \\ \mu(\omega) &= diag\left[\; \mu_1(\omega) \quad \mu_2(\omega) \quad \mu_3(\omega) \; \right] \\ &= \begin{bmatrix} \frac{\sin(r_1 \pi)}{\pi} \omega^{-r_1} & 0 & 0 \\ 0 & \frac{\sin(r_2 \pi)}{\pi} \omega^{-r_2} & 0 \\ 0 & 0 & \delta(\omega) \end{bmatrix} \end{aligned} \tag{18}$$

After the equivalent transformation, the fractional-order model is transformed into a continuous frequency distributed state model. In order to obtain the convergence of the designed non-linear observer based on an RBF neural network, the Lyapunov function $V(t)$ is chosen as follows:

$$V(t) = \int_0^\infty z^T(\omega, t) \mu(\omega) P z(\omega, t) d\omega + \frac{1}{2} \widetilde{W} \Gamma^{-1} \widetilde{W} \tag{19}$$

Then, Inequation (20) can be obtained.

$$
\begin{aligned}
\dot{V}(t) &= \int_0^\infty \left[\frac{\partial z^T(\omega,t)}{\partial t} \mu(\omega) P z(\omega,t) + z^T(\omega,t)\mu(\omega) P \frac{\partial z(\omega,t)}{\partial t} \right] d\omega + \widetilde{W}\Gamma^{-1}\dot{\widetilde{W}} \\
&= \int_0^\infty \left[-\omega z(\omega,t) + \widetilde{A}_{cl}\widetilde{x}(t) + (\widetilde{B} - L\widetilde{D})I_B(t) + LF(t) \right. \\
&\quad \left. + L(\varphi(x,I_B) - \hat{\varphi}(\hat{x},I_B)) \right]^T \mu(\omega) P z(\omega,t) d\omega \\
&\quad + \int_0^\infty z^T(\omega,t)\mu(\omega)P[-\omega z(\omega,t) + \widetilde{A}_{cl}\widetilde{x}(t) + (\widetilde{B} - L\widetilde{D})I_B(t) \\
&\quad + LF(t) + L(\varphi(x,I_B) - \hat{\varphi}(\hat{x},I_B))]d\omega - \widetilde{W}K_\omega W \\
&\leq \int_0^\infty [\widetilde{x}^T(t)\widetilde{A}_{cl}^T\mu(\omega)Pz(\omega,t) + (\widetilde{B} - L\widetilde{D})I_B(t)\mu(\omega)Pz(\omega,t) \\
&\quad + F^T(t)L^T\mu(\omega)Pz(\omega,t) + (\varphi(x,I_B) - \hat{\varphi}(\hat{x},I_B))^T L^T \mu(\omega)Pz(\omega,t)]d\omega \\
&\quad + \int_0^\infty [z^T(\omega,t)\mu(\omega)P\widetilde{A}_{cl}\widetilde{x}(t) + z^T(\omega,t)\mu(\omega)PL(\varphi(x,I_B) - \hat{\varphi}(\hat{x},I_B)) \\
&\quad + z^T(\omega,t)\mu(\omega)P(\widetilde{B} - L\widetilde{D})I_B(t) + z^T(\omega,t)\mu(\omega)PLF(t)]d\omega - \widetilde{W}K_\omega W \\
&= \widetilde{x}^T(t)(\widetilde{A}_{cl}^T P + P\widetilde{A}_{cl})\widetilde{x}(t) + 2\widetilde{x}^T(t)P(\widetilde{B} - L\widetilde{D})I_B(t) + 2\widetilde{x}^T(t)PLF(t) \\
&\quad + 2\widetilde{x}^T(t)PL(\varphi(x,I_B) - \hat{\varphi}(\hat{x},I_B)) - \widetilde{W}K_\omega W
\end{aligned}
\tag{20}
$$

Assume $P = \mathrm{diag}\{\ 1/2p_1 \quad 1/2p_2 \quad 1/2p_3\ \}$ and $p_1, p_2, p_3 > 0$ then:

$$
\begin{aligned}
\dot{V}(t) &\leq -p_1(m - l_1)\widetilde{x}_1^2 + \left|(p_1 l_1 + p_2 l_2)\widetilde{x}_1\widetilde{x}_2\right| + \left|p_3 l_3 \widetilde{x}_1\widetilde{x}_3\right| - p_2(n - l_2)\widetilde{x}_2^2 \\
&\quad + \left|p_3 l_3 \widetilde{x}_2\widetilde{x}_3\right| - p_3(a - \gamma l_3)\widetilde{x}_3^2 + \left|p_1[(g_1(x_3) - g_1(\hat{x}_3)) + l_1(g_3(\hat{x}_3) - g_3(x_3))]I_B\widetilde{x}_1\right| \\
&\quad + \left|p_2[(g_2(x_3) - g_2(\hat{x}_3)) + l_2(g_3(\hat{x}_3) - g_3(x_3))]I_B\widetilde{x}_2\right| \\
&\quad + \left|p_3[l_3(g_3(\hat{x}_3) - g_3(x_3))]I_B\widetilde{x}_3\right| + \left|p_3 l_3 \varepsilon_M \widetilde{x}_3\right| - \widetilde{W}K_\omega W
\end{aligned}
\tag{21}
$$

Because $g_i(\cdot)(1 \leq i \leq 3)$, $x_i(\cdot)(1 \leq i \leq 3)$, $I_B(t)$ are bounded, Inequation (22) can be obtained.

$$
\begin{aligned}
|p_1[(g_1(x_3) - g_1(\hat{x}_3)) + l_1(g_3(\hat{x}_3) - g_3(x_3))]I_B| &\leq M_1 \\
|p_2[(g_2(x_3) - g_2(\hat{x}_3)) + l_2(g_3(\hat{x}_3) - g_3(x_3))]I_B| &\leq M_2 \\
|p_3[l_3(g_3(\hat{x}_3) - g_3(x_3))]I_B| &\leq M_3
\end{aligned}
\tag{22}
$$

Assume $(l_1 p_1 + l_2 p_2) = N$; Inequation (23) can be obtained.

$$
\begin{aligned}
\dot{V}(t) &\leq -p_1(m - l_1)\widetilde{x}_1^2 + \left|N\widetilde{x}_1\widetilde{x}_2\right| + \left|p_3 l_3 \widetilde{x}_1\widetilde{x}_3\right| - p_2(n - l_2)\widetilde{x}_2^2 + \left|p_3 l_3 \widetilde{x}_2\widetilde{x}_3\right| \\
&\quad - p_3(a - \gamma l_3)\widetilde{x}_3^2 + p_1 M_1\left|\widetilde{x}_1\right| + p_2 M_2\left|\widetilde{x}_2\right| + p_3 M_3\left|\widetilde{x}_3\right| + \left|p_3 l_3 \varepsilon_M \widetilde{x}_3\right| - \widetilde{W}K_\omega W
\end{aligned}
\tag{23}
$$

Then, the inequation can be obtained as shown in Inequations (24) and (25).

$$
\begin{aligned}
(1) &\ -p_1(m - l_1)\widetilde{x}_1^2 + p_1 M_1|\widetilde{x}_1| + |N\widetilde{x}_1\widetilde{x}_2| \leq -\frac{p_1(m-l_1)}{3}\widetilde{x}_1^2 + \frac{3p_1 M_1^2}{(m-l_1)} + \frac{3N^2}{p_1(M_1 - l_1)}\widetilde{x}_2^2 \\[4pt]
(2) &\ -p_2(n - l_1)\widetilde{x}_2^2 + |p_2 M_2\widetilde{x}_2| \leq -\frac{p_2(n-l_2)}{2}\widetilde{x}_2^2 + \frac{2p_2 M_2^2}{(n-l_2)} \\[4pt]
(3) &\ -p_3(a - \gamma l_3)\widetilde{x}_3^2 + |p_3 l_3 \widetilde{x}_2\widetilde{x}_3| + |p_3 l_3 \widetilde{x}_1\widetilde{x}_3| + |p_3 M_3\widetilde{x}_2| + p_3 l_3 \varepsilon_M|\widetilde{x}_3| \leq -\frac{p_3(a - \gamma l_3)}{4}\widetilde{x}_3^2 \\
&\ + \frac{4p_3 l_3^2 \widetilde{x}_1^2}{(a - \gamma l_3)} + \frac{4p_3 l_3^2 \widetilde{x}_2^2}{(a - \gamma l_3)} + \frac{4p_3(M_3 + l_3 \varepsilon_M)^2}{(a - \gamma l_3)}
\end{aligned}
\tag{24}
$$

$$
-\widetilde{W}K_\omega W = -\widetilde{W}K_\omega\left(\widetilde{W} + W^*\right) \leq -\frac{\|K_\omega\|\left\|\widetilde{W}\right\|^2}{2} + \frac{\|K_\omega\|\|W_M\|^2}{2}
\tag{25}
$$

According to Inequations (24) and (25), Inequation (23) can be rewritten as Inequation (26).

$$
\dot{V}(t) \leq \alpha_1 \widetilde{x}_1{}^2 + \alpha_2 \widetilde{x}_2{}^2 + \alpha_3 \widetilde{x}_3{}^2 - \frac{\|K_\omega\|\left\|\widetilde{W}\right\|^2}{2} + M
\tag{26}
$$

where

$$a_1 = \frac{-p_1(m-l_1)}{3} + \frac{4p_3 l_3^2}{(a-\gamma l_3)}$$
$$a_2 = -\frac{p_2(n-l_2)}{2} + \frac{4p_3 l_3^2}{(a-\gamma l_3)} + \frac{3N^2}{p_1(m-l_1)}$$
$$a_3 = \frac{-p_3(a-\gamma l_3)}{4}$$
$$M = \frac{3p_1^2 M_1^2}{(m-l_1)} + \frac{2p_2 M_2^2}{(n-l_2)} + \frac{4p_3(M_3+l_3\varepsilon_M)^2}{(a-\gamma l_3)} + \frac{\|K_\omega\|\|W_M\|^2}{2}$$

(27)

According to the previous analysis, if Equation (14) is satisfied by l_1, l_2, l_3, and α_1, α_2, α_3 can be guaranteed to be negative. Since the initial value of Lyapunov function $V(0)$ is bounded, the solution of the estimation error system is also uniformly bounded according to the boundedness analysis. The bounds of *SOC* estimation error are satisfied with $\{x_3(t) \leq \sqrt{M/\lambda}\}$, in which λ is a constant satisfied with $0 < \lambda < p_3(a - \gamma l_3)/4$. If p_3 is selected as large, λ can be selected large enough, and then the bounds of *SOC* estimation error can be arbitrarily small. Therefore, a non-linear observer based on an RBF neural network can accurately estimate the *SOC* of batteries. $\square$

4. Experiment and Simulation Analysis

The test system was built with a PXI system developed by the National Instruments (NI). The electrical schematic diagram of the test system is shown in Figure 3, and the battery test bench was built as shown in Figure 4. The hardware platform of the battery test system was composed of four parts: an industrial computer, charging and discharging instruments, a parameter measurement device and a thermostatic humidity test box, and the software of the battery test system was designed through the LabVIEW development environment. The environmental temperature was simulated by adjusting the internal temperature of the given thermostatic humidity box. The charging and discharging instruments were controlled by a program, the charging and discharging currents of the battery were set during the experiment, and the performance test of the power battery in different environments was completed. In the process of the experiment, the current voltage and temperature of the battery were collected in real time using a high-precision voltage current and temperature sensor, and the real-time data collected were saved and exported to the table using the SQL database linked to the experimental platform.

The equipment and software list in Figure 3 is shown in Table 1.

Table 1. The equipment and software list.

No.	Name of Equipment or Software	Model or Version	Functions and Performance
1	DC Power supply	IT6533A	160 V, 120 A, 6 kW
2	Electronic Load	IT8830	120 V, 500 A, 10 kW
3	USB-DAQ card	NI USB-6216	16 bit, 400 kS/s, Isolated Multifunciton DAQ
4	USB-DMM	NI USB-4065	$6_{1/2}$-Digit, ± 300 V, 0.025%
5	Heat Test Chamber	WTH-150-40-880	$-45\,^{\circ}$C~105 $^{\circ}$C, 99 Rh%
6	Labview	2019 SP1	
7	Matlab	2019b	
8	GPIB Card	NI USB-GPIB	
9	USB-CAN Card	Peak CAN-USB	
10	High Performance Voltage Isolated Interface	HVIS-400	Isolated Voltage: 500 V, 0.01%
11	High Performance Current Isolated Interface	HIIS-100	Isolated Voltage: 500 V, 0.01%
12	PXI Controller	NI PXIe-8840	Intel Core i7, 8 GB/s
13	PXI Chassis	NI PXIe-1062	8 slots, PXI Express Chassis

The test system, composed of a heat test chamber, DC power supply, electronic load and PXI system, is shown in Figure 4.

The UI of the lithium battery test system based on Labview is shown in Figure 5. The characteristic test for lithium-ion power battery could be selected on the main interface in real time, and the current size of the charger and discharger could be set. The temperature, charging and discharging current, voltage and other parameters of the battery could be displayed on the interface in real time.

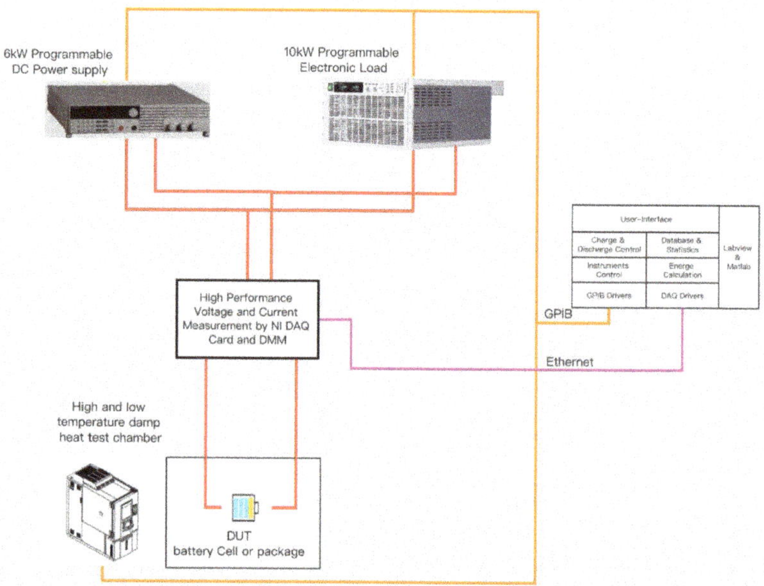

Figure 3. Architecture diagram of battery characteristic test platform.

Figure 4. Photos of battery characteristic test platform.

The open-circuit voltage test and HPPC test of batteries were carried out using the test and verify system. The parameters of the above model could be obtained by the least squares method. Then, these curves were fitted to balance the accuracy and complexity using an empirical equation composed of exponential and polynomial function. Figure 6a,b show the non-linear relationship between OCV and SOC of batteries and the curves between parameters (R_0, R_1, C_1, R_2 and C_2) and SOC in the equivalent circuit of batteries. The red dot is the data recorded in the test, and the blue line is the corresponding fitting curve for the actual data obtained from the experiment. The weight vector rule of the neural network is $K_\omega = diag(0.01, 0.01, 0.01, 0.01, 0.01, 0.01, 0.01, 0.01)$ and $\Gamma = diag(1, 1, 1, 1, 11, 1, 1, 1, 1)$.

Figure 6a shows the curves of OCV and SOC obtained using the standard OCV test method. Figure 6b shows the measured values and fitting curves of the equivalent circuit parameters of the battery.

The battery test platform mentioned above was used to carry out three different dynamic operating conditions experiments to verify the accuracy and effectiveness of the proposed RBF neural network, non-linear, fractional-order observer. Since high-precision measuring instruments were used when building the battery test platform, the current data collected by the device had high precision. The accurate SOC could be obtained as the real reference value of SOC using the current-time integral method.

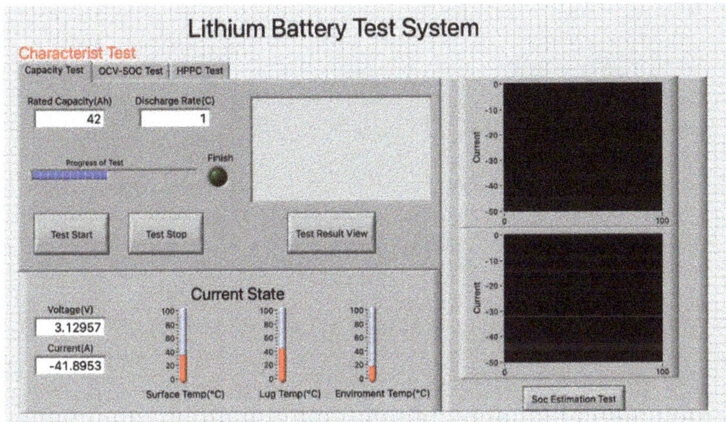

Figure 5. Battery test system interface.

Dynamic operation test 1: Leave the fully charged battery in the incubator for 24 h at the laboratory temperature to ensure that the temperature of the entire battery is even. Use a constant current of 1C to discharge the battery at a capacity of 10% (to avoid overvoltage conditions during the subsequent dynamic operating experiment). Dynamic operation test 1 is performed within the range of SOC of interest (90% to 10% SOC). The experimental current image of Dynamic operation test 1 is shown in Figure 7.

The measured actual end voltage for reference and the model estimated end voltage curves and their local amplification diagrams are shown in Figures 8 and 9, respectively. The local amplification diagrams shown in Figure 9 show that the measured actual end voltage curves for reference and the curves estimated by the model basically coincide. The RMSE of the real and model results was 5.97 mV under dynamic condition test 1. The local amplification curve shown in Figure 9 shows more clearly that the identified model reflects the dynamic characteristics of the battery with high accuracy, and the proposed circuit model captures the battery performance well. The accuracy of battery SOC estimation depends on whether the battery voltage estimated by the model is accurate, which proves that the proposed model can more effectively and accurately estimate the battery status, and the corresponding voltage error curve is shown in Figure 10.

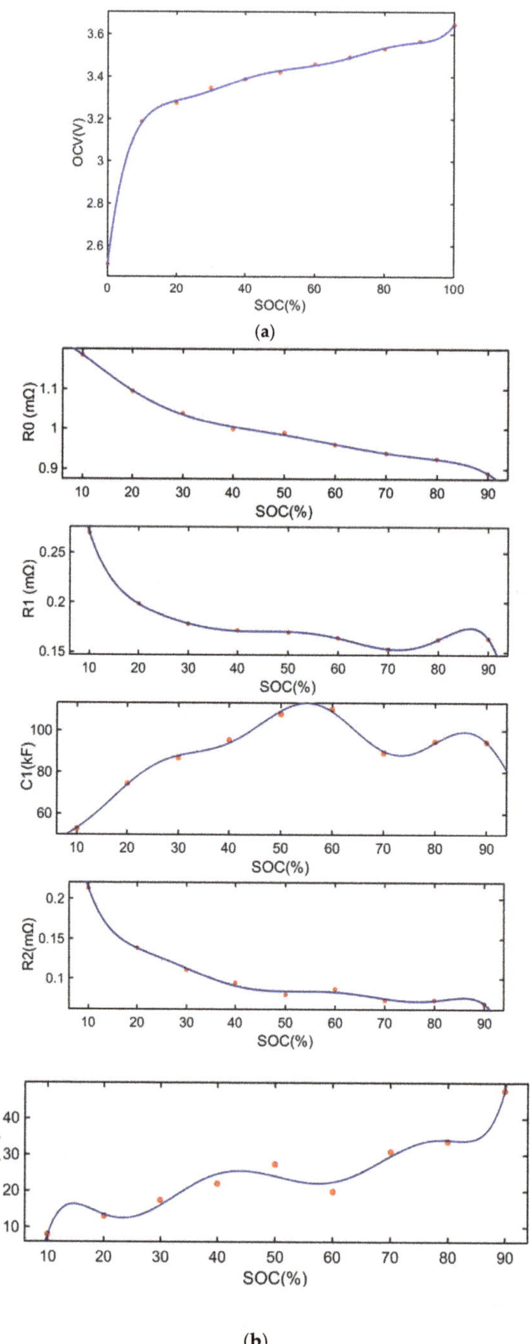

Figure 6. (**a**) Relationship between *OCV* and *SOC* of battery. (**b**) Parameters of battery equivalent circuit model and its fitting curve.

The results of estimating *SOC* using the proposed RBF neural network non-linear observer and EKF method are shown in Figure 11. It can be seen that the proposed RBF neural network non-linear observer accurately estimated the *SOC*. The error between the true value and the estimated value decreased gradually. The final estimated curve converged gradually near the true value curve, and the *SOC* error was limited to a very narrow error range. It was related to accurately modeling and accurately estimating the battery end voltage, which was used to correct the estimated *SOC*. Furthermore, this proved that the identified fractional-order electrochemical impedance model could accurately capture the battery dynamics and accurately estimate the battery end voltage, thus enabling the designed RBF neural network, non-linear, fractional-order observer to accurately and effectively estimate the *SOC* and power in batteries.

In order to compare the error of *SOC* estimation between the two methods, the error of *SOC* state estimation is also shown in Figure 12. The curve of the proposed RBF neural network non-linear observer *SOC* estimation error was nearly zero, and *SOC* error mainly fluctuated within the range of +0.5%. Conversely, *SOC* estimation error based on EKF method varied greatly, and the error range was large. The validity of the proposed RBF neural network non-linear observer is further proved.

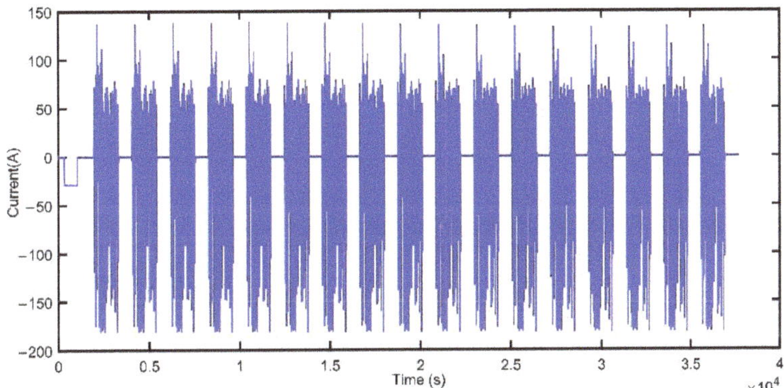

Figure 7. Current of dynamic condition test 1 experiment.

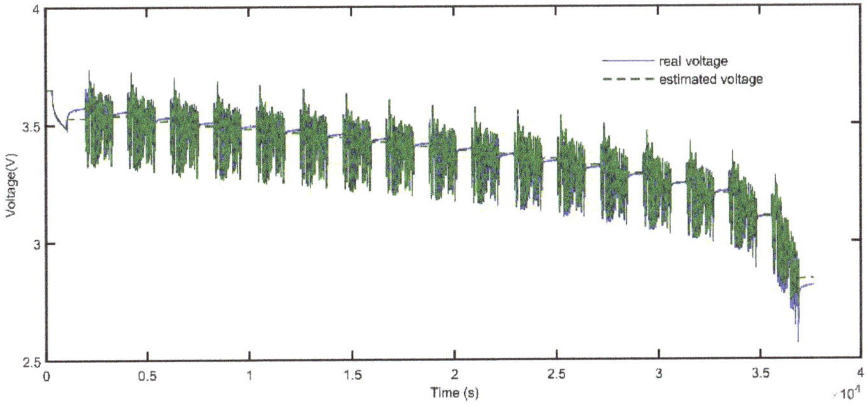

Figure 8. Comparison of real and estimated voltage of test 1 under dynamic condition.

In practical engineering applications, the ability to correct *SOC* is necessary for a good *SOC* estimation algorithm. Therefore, in the case of dynamic condition 1, the strategy of random *SOC* initial value was adopted to verify the *SOC* correction ability of the proposed algorithm. As shown in Figures 13 and 14, when the initial *SOC* value deviated from the real value, the RBF neural network non-linear observer could quickly converge to the true value in a short time and keep the error below 2%, while the EKF method converged to the true value almost at the end of the test. The convergence speed of the proposed RBF neural network non-linear observer is further proved.

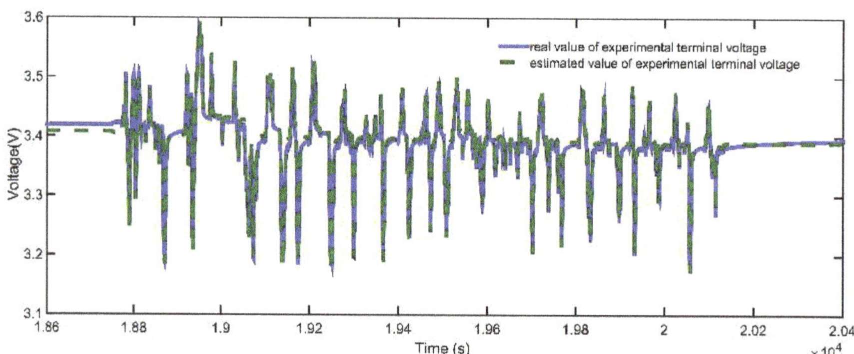

Figure 9. Comparison of real value and estimated value of experimental terminal voltage in dynamic condition test 1.

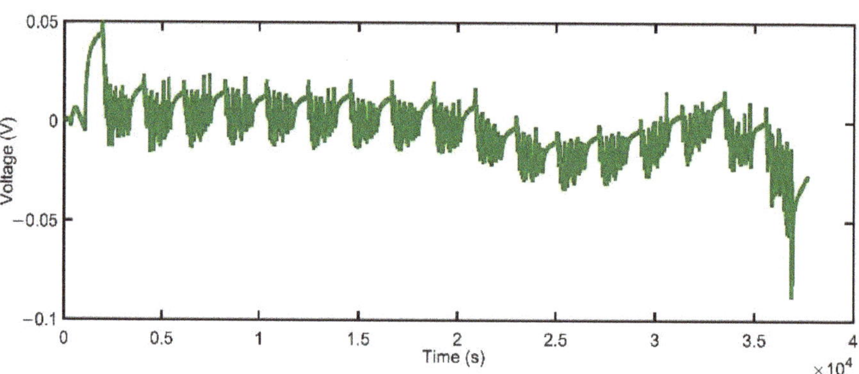

Figure 10. Voltage estimation error of experiment terminal in dynamic condition test 1.

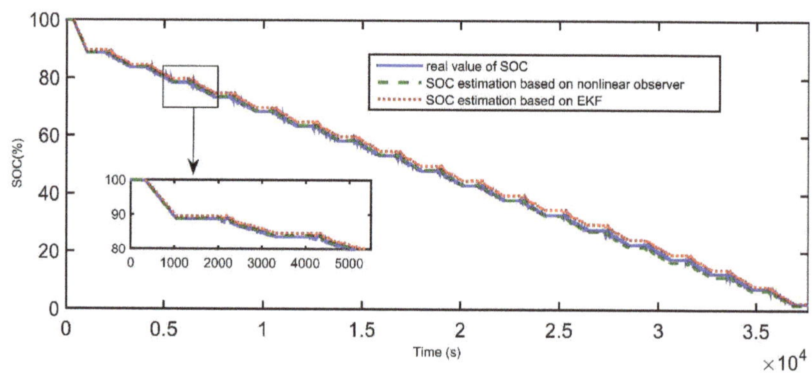

Figure 11. Comparison of *SOC* estimation based on non-linear observer and EKF in dynamic condition test 1.

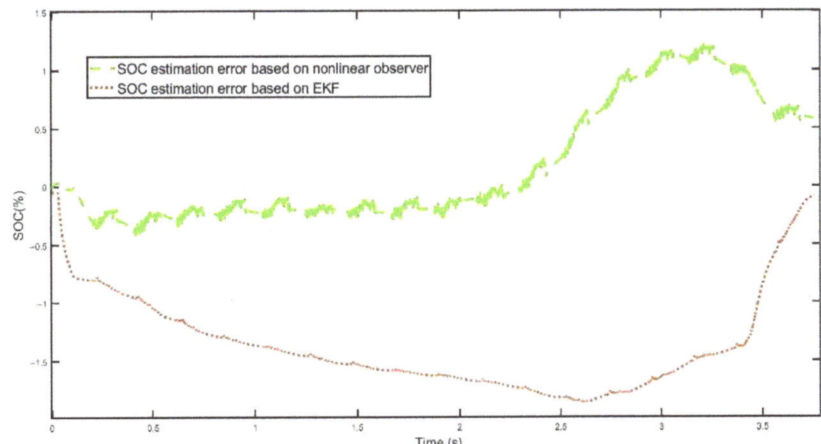

Figure 12. Comparison of *SOC* estimation error based on non-linear observer and EKF in dynamic condition test 1.

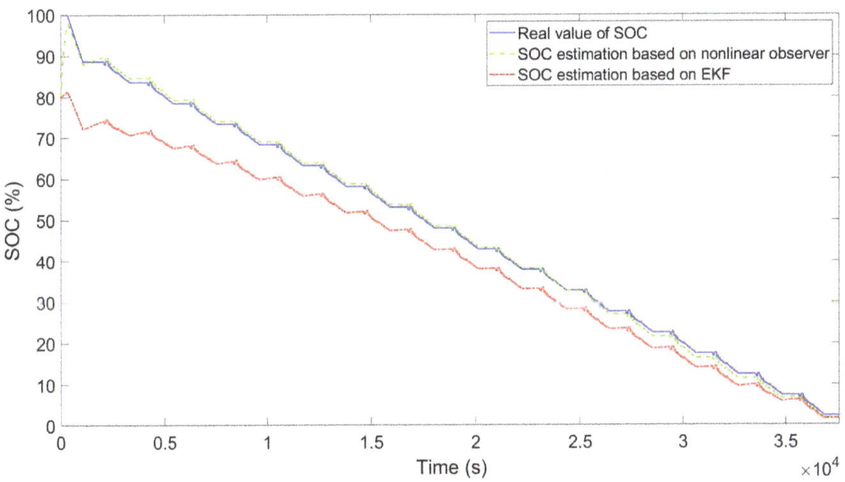

Figure 13. Comparison of *SOC* estimation based on non-linear observer and EKF in dynamic condition test 1 under random *SOC* initial value.

Remark 3. *The effects of the number of ganglion points on the performance and calculation of SOC estimation for lithium-ion power batteries based on an RBF neural network non-linear observer were studied through comparative experiments and simulations. More specifically, two, five, seven, eleven and fifteen lithium-ion power batteries were compared to estimate their SOC performance and load using the experimental data of battery current and end voltage collected. The performance comparison of the RBF neural network non-linear observers with different ganglion points for estimating SOC is shown in* Figure 15. *The running time for solving the non-linear observer problem of RBF neural networks with different number of ganglion points is shown in* Figure 16 *to represent the computational burden of different ganglion points. From the comparison results, the neural network non-linear observer used in the experiment with seven ganglion points has good performance under a reasonable amount of calculation.*

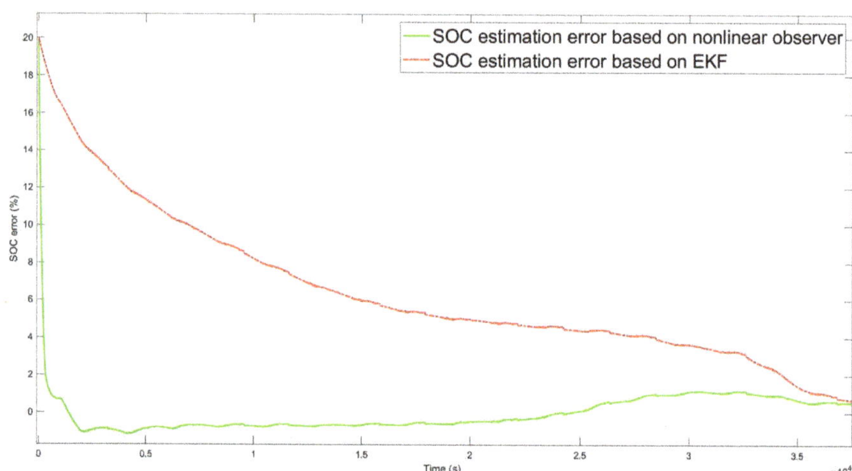

Figure 14. Comparison of *SOC* estimation based on non-linear observer and EKF in dynamic condition test 1 under random *SOC* initial value.

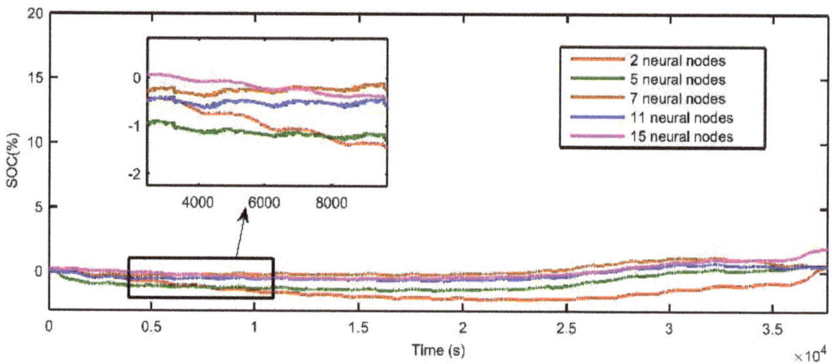

Figure 15. *SOC* estimation error of non-linear observers with different numbers of neural nodes.

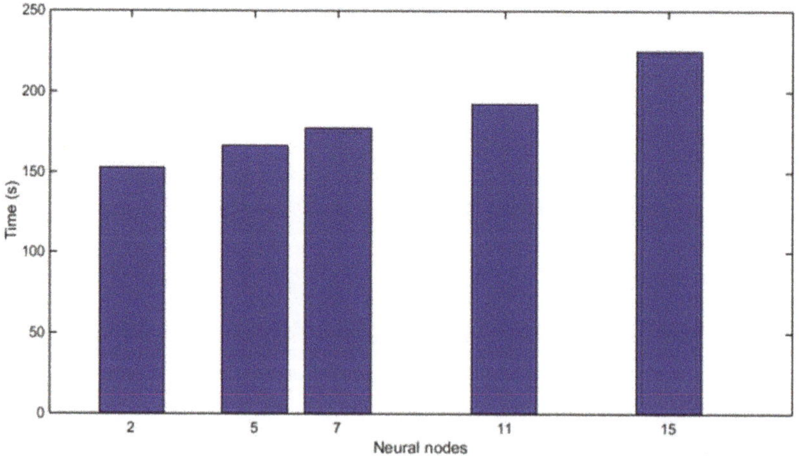

Figure 16. Time consuming of solving non-linear observers with different numbers of neural nodes.

Dynamic operation test 2: Leave the fully charged battery in the incubator for 24 h at the laboratory temperature to ensure that the temperature of the whole battery is even. The battery is charged to full capacity with a constant current of 1C, and then tested under dynamic condition 2. When fully charged, the battery is discharged with high current for a period, and then charged and discharged with the current changing with dynamic fluctuations. The current image of dynamic condition test 2 is shown in Figure 17.

The measured actual end voltage used for reference and the model estimated end voltage curve are shown in Figure 18. It can be seen that the measured actual end voltage curve used for reference and the curve estimated by the model did not coincide as in Working Condition 1. The main reason for this is that the voltage fluctuation in the second half of the test was so small that the model did not track the voltage change in the power battery very accurately. The corresponding voltage error curves are shown in Figure 19. The RMSE of the true and model results was 11.7 mV under the dynamic condition test 2. This indicates that the model was less accurate under very small voltage fluctuations, but the accuracy of the estimation was still high enough.

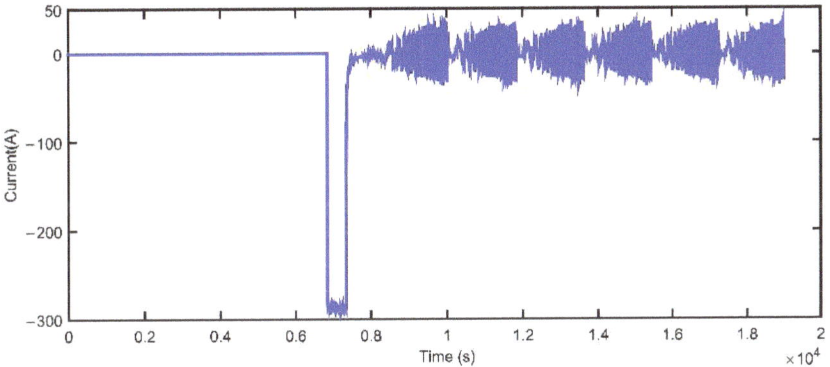

Figure 17. Dynamic condition test 2 experimental current.

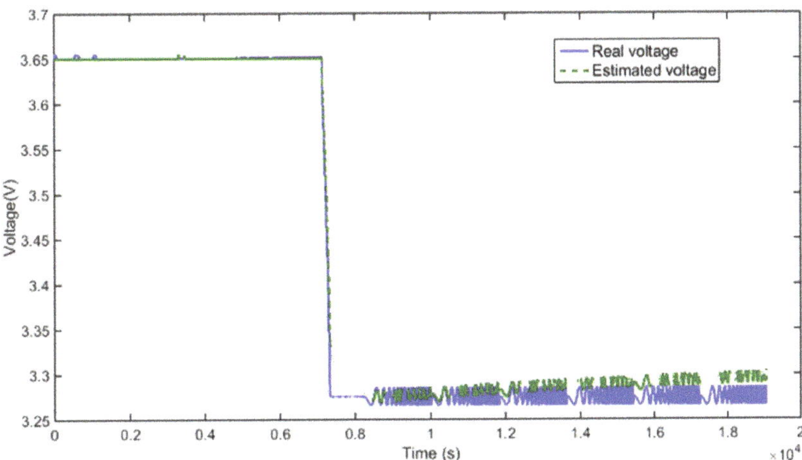

Figure 18. Comparison of real and estimated voltage of test 2 under dynamic condition.

The results of estimating *SOC* using the proposed RBF neural network non-linear observer and the EKF method are shown in Figure 20. The error of estimating *SOC* state using the proposed RBF neural network non-linear observer and the EKF method is shown in Figure 21. It can be seen that the proposed RBF neural network non-linear observer accurately estimated the *SOC* of power batteries, and the estimated *SOC* error decreased gradually. However, due to the small fluctuation of voltage in the second half of the test period, the *SOC* estimation error did not converge to a range gradually because the model did not track the change in power battery end voltage very accurately. However, it can be seen from Figure 21 that the *SOC* estimation curve of the proposed non-linear observer of the RBF neural network was still very close to the real value of *SOC*. On the contrary, although the *SOC* estimation curve obtained using the EKF method was closer to the true value in the early stage, the error increased during the back-end test process. The proposed RBF neural network non-linear observer was more accurate than the EKF method.

Dynamic operation test 3: Leave the fully charged battery in the incubator for 26 h at the laboratory temperature to ensure that the temperature of the whole battery is even. Charge the battery to its full capacity using a constant current of 1C and discharge it to zero power battery capacity. Test under dynamic condition 3, charge the battery with a high current for a period after discharging, and then charge and discharge with dynamic fluctuating current. The image of the current under dynamic condition test 2 is shown in Figure 22.

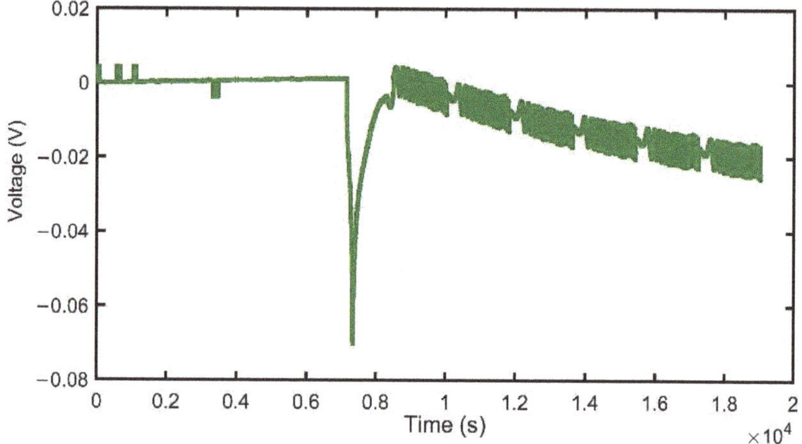

Figure 19. Voltage estimation error of test 2 under dynamic condition.

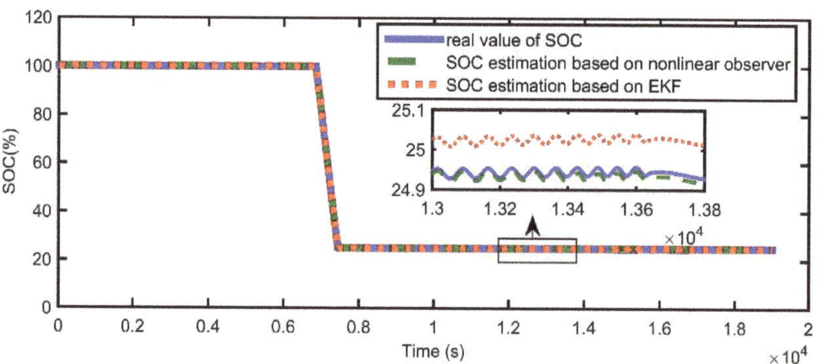

Figure 20. Comparison of *SOC* estimation based on non-linear observer and EKF in dynamic condition test 2.

The measured actual end voltage for the reference and the model-estimated end voltage curve are shown in Figure 23. The measured actual end voltage curve coincided with the model estimated curve. The corresponding voltage error curve is shown in

Figure 24. The RMSE of the real and model results was 9.69 mV under dynamic condition test 3, which indicates that the proposed electrochemical impedance model could accurately estimate the terminal voltage of the power battery.

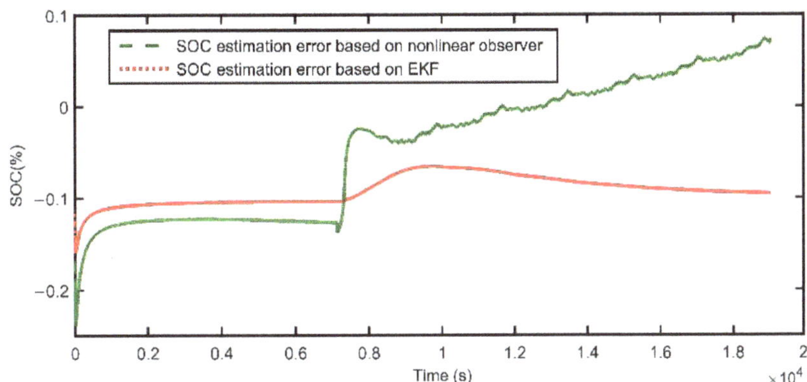

Figure 21. Comparison of *SOC* estimation error based on non-linear observer and EKF in dynamic condition test 2.

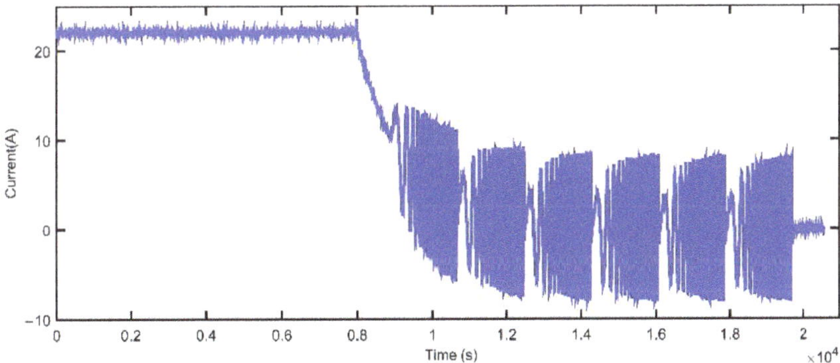

Figure 22. Dynamic condition test 3 experimental current.

The results of estimating *SOC* using the proposed RBF neural network non-linear observer and the EKF method are shown in Figure 25. It can be seen that the proposed RBF neural network non-linear observer accurately estimated the *SOC*, and the error between the true value and the estimated value was small. The *SOC* state estimation error estimated using the proposed RBF neural network non-linear observer and the EKF method is shown in Figure 26. It can be observed that the *SOC* estimation error curve of the proposed RBF neural network non-linear observer was closer to the transverse coordinate axis. Although the *SOC* error curve estimated using the EKF method decreased gradually at the end, the *SOC* estimation error was larger in the middle of the curve. The accuracy of the proposed non-linear observer for RBF neural networks is further demonstrated.

To further show the accuracy of the proposed RBF neural network non-linear observer in estimating the *SOC* of the EKF method, Table 2 gives the RMSE results of estimating the *SOC* using the proposed RBF neural network non-linear observer and the EKF method under three dynamic operating conditions tests. Figures 12, 21 and 26 and Table 2 show that the RBF non-linear observer designed has higher estimation accuracy than the EKF.

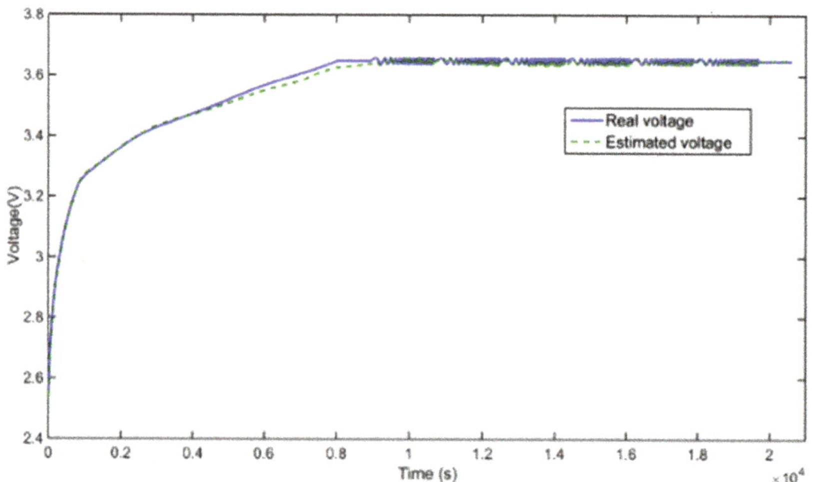

Figure 23. Comparison of real and estimated voltage of test 3 under dynamic condition.

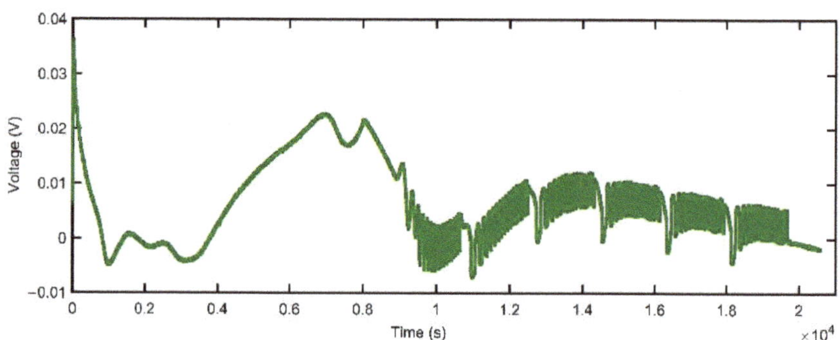

Figure 24. Voltage estimation error of test 3 under dynamic condition.

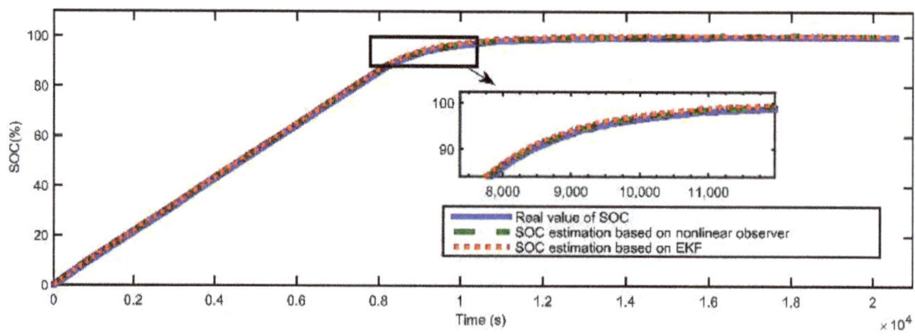

Figure 25. Comparison of *SOC* estimation based on non-linear observer and EKF in dynamic condition test 3.

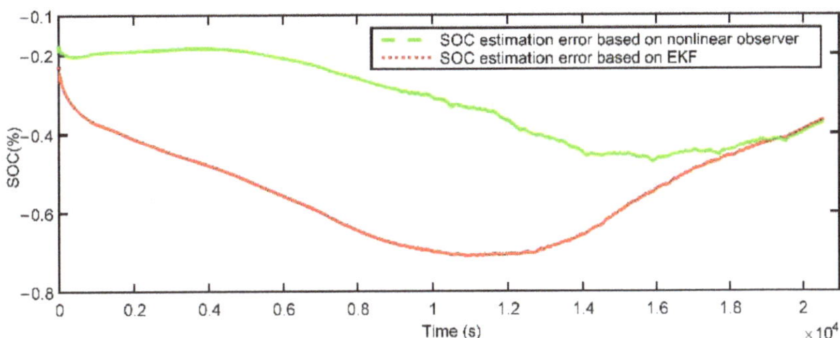

Figure 26. Comparison of *SOC* estimation error based on non-linear observer and EKF in dynamic condition test 3.

Table 2. RMSE of *SOC* estimation error based on RBF neural network non-linear observer and EKF method.

	Working Condition 1	**Working Condition 2**	**Working Condition 3**
RBF neural network non-linear observer	0.53%	0.24%	0.33%
EKF method	1.42%	1.70%	0.75%

The comparison of the accuracy of the EKF method with the RBF neural network non-linear observer proposed under different dynamic test conditions is given. The following are the simulation results of voltage and *SOC* estimation using the proposed RBF neural network non-linear observer under dynamic test condition 1 under integer- and fractional-order models.

Figure 27 shows the estimation of battery end voltage using the integer-order model and the proposed electrochemical impedance fractional-order model, the actual measured end voltage, and the comparison of the estimated end voltage curves between the two models. Figure 28 shows the local amplification comparison of the two models, and the corresponding voltage error curves are shown in Figure 29. The local magnification diagram shown in Figure 28 shows that the measured actual end voltage curve used for reference coincides basically with the curve estimated based on the fractional model of electrochemical impedance, while the voltage estimate based on the integer-order model has a greater error than that based on the fractional-order model. The RMSE based on fractional-order model voltage estimation was 5.97 mV, and 13.7 mV based on integer-order model voltage estimation. This shows that the proposed fractional-order model of electrochemical impedance for power batteries can clearly reflect its dynamic characteristics and battery performance.

The proposed non-linear observer of the RBF neural network was used to estimate *SOC* based on the fractional- and integer-order models of electrochemical impedance proposed. The estimation curve of *SOC* is shown in Figure 30. From Figure 30, it can be seen that the *SOC* estimation curve of the fractional electrochemical impedance model proposed basically coincided with the true value curve, and the estimation result was better than that based on the integer-order model. The RMSE of *SOC* based on the fractional model of electrochemical impedance was 0.53%, and that of *SOC* based on the integer model was 1.26%. Using the proposed RBF neural network non-linear observer to estimate *SOC* based on the proposed fractional and integer-order models, the results show that the estimated *SOC* error based on the fractional-order model of electrochemical impedance was between (+) 0.5% and that based on the integer-order model was between 0.5% and 2%. The proposed fractional model of electrochemical impedance can more accurately estimate the *SOC* of power batteries.

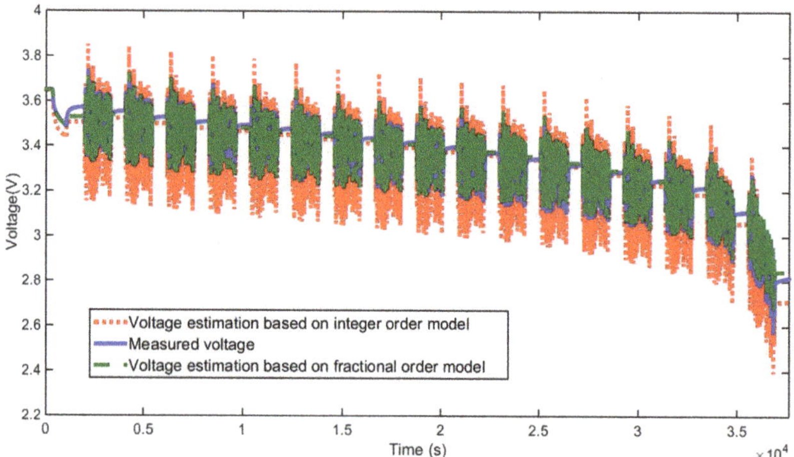

Figure 27. Comparison of voltage estimation based on fractional-order model and integer-order model.

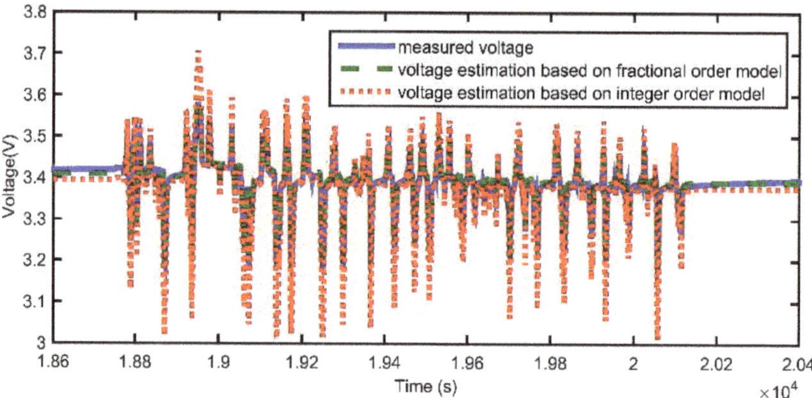

Figure 28. Comparison and enlargement of voltage estimation based on fractional-order model and integer-order model.

The following conclusions can be drawn from the above experimental simulation results in Figures 11, 12, 20, 21, 25, 26, 30 and 31 and Table 1:

(1) From Figures 30 and 31, the proposed non-linear observer of RBF neural network has higher accuracy in estimating *SOC* of the fractional-order model of electrochemical impedance than that of the integer-order model, because the proposed fractional-order model of electrochemical impedance has smaller modeling error, more accurate estimation of end voltage of power batteries and less error in estimating the *SOC* state of batteries.

(2) For power battery *SOC* estimation, the proposed RBF-neural-network-based non-linear observer has smaller error and higher accuracy than the EKF method. From Figures 11, 20 and 25, the curve of *SOC* estimation using the EKF method is smoother than that of the proposed RBF neural network non-linear observer. The reason for these two phenomena is that the observer-based method is superior to the filter in estimating the system output. The observer-based non-linear observer based on the RBF neural network can quickly catch the change when the experimental

current changes significantly between positive and negative modes in three dynamic operating conditions. The convergence rate of state estimation is then adjusted.

(3) From Figures 12, 21 and 26, it can be seen that the proposed non-linear observer method for RBF neural networks converges faster than the EKF in *SOC* state estimation. The RMSE of the *SOC* estimation errors for the two methods in Table 1 further demonstrates that the proposed non-linear observer method for RBF neural networks is superior to the EKF method.

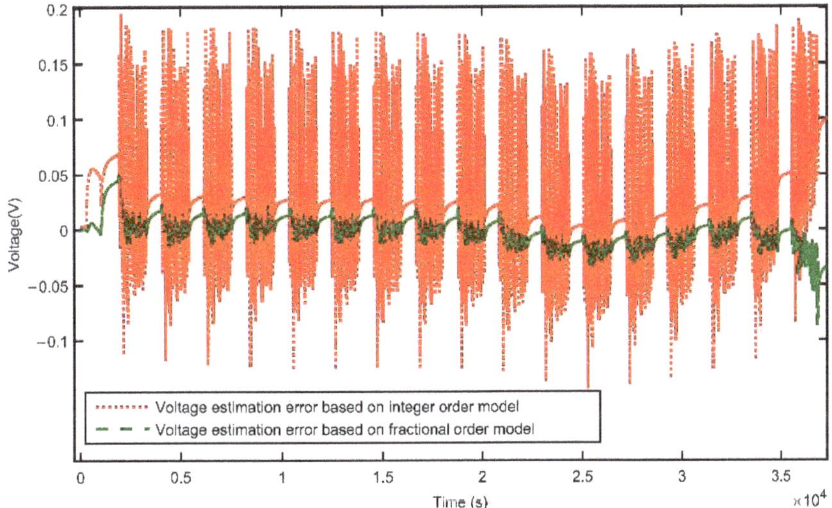

Figure 29. Comparison of voltage estimation error based on fractional-order model and integer-order model.

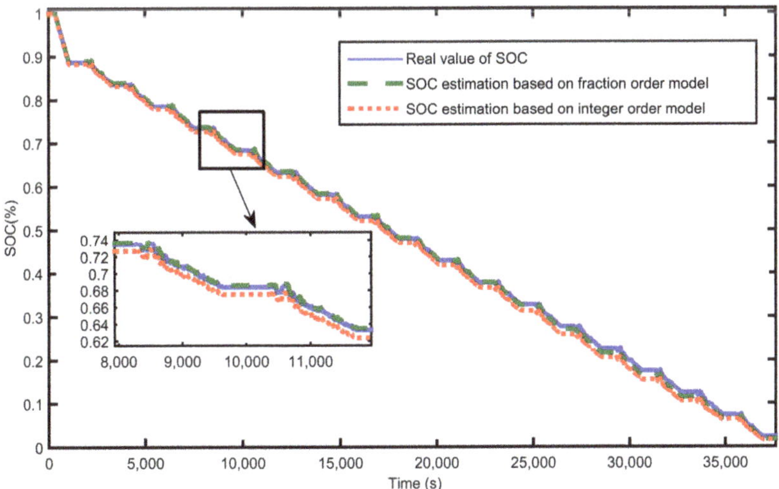

Figure 30. Comparison of *SOC* estimates based on fractional-order model and integer-order model.

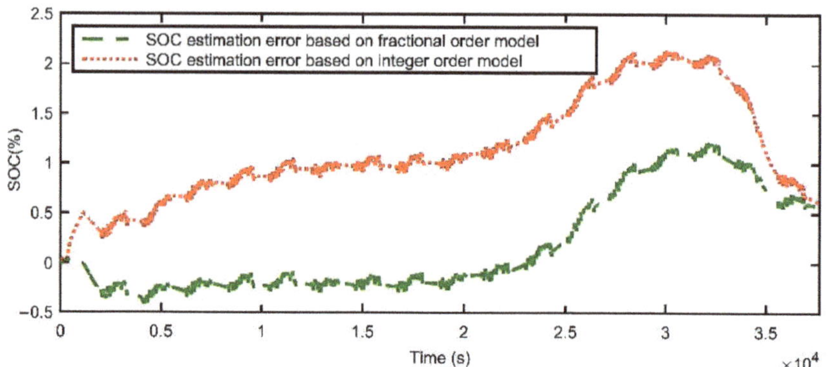

Figure 31. Comparison of *SOC* estimation error based on fractional-order model and integer-order model.

Based on the above analysis, it is known that the fractional model of electrochemical impedance proposed has less modeling error and higher estimation accuracy than the integer model. For the fractional-order model of electrochemical impedance proposed, the proposed RBF neural network non-linear observer method is more effective and more accurate than the EKF method. The proposed RBF neural network non-linear observer method can run in real time and reliably and accurately estimate the end voltage and *SOC* state of power batteries. In summary, the fractional-order model of electrochemical impedance and the RBF neural network non-linear observer method proposed not only take into account the characteristics of the battery itself, but also show good performance.

5. Conclusions

This paper designed a method for the state of charge (SOC) estimation of lithium-ion batteries based on an electrochemical impedance equivalent circuit model with a controlled source. A non-linear observer was designed to estimate the *SOC* of batteries by using an RBF neural network to estimate the uncertainty of batteries online. The estimation error of the non-linear observer based on an RBF neural network was proved to be ultimately bounded by Lyapunov stability analysis. The results of simulations and experiments in this paper are as follows: compared with the EKF, the average error of *SOC* measurement with the non-linear observer was reduced by 50%, and the error of the optimal experimental result was reduced by 70%. Compared with the integer-order model, the measurement error of the fractional-order model was reduced by 40% on average, and the error of the optimal experimental result was reduced by 60%. The experimental and simulation results show the accuracy and superiority of the proposed method.

Author Contributions: N.C.: Conceptualization, Writing—Review and Editing, Supervision. X.Z.: Methodology, Software, Writing—Review and Editing. J.C.: Validation, Methodology, Writing—Review and Editing. X.X.: Writing—Review and Editing, Supervision, Project Administration. P.Z.: Methodology, Software. W.G.: Conceptualization, Project administration. All authors have read and agreed to the published version of the manuscript.

Funding: This work was supported in part by the Key Program of National Natural Science Foundation of China (62033014), in part by the Application Projects of Integrated Standardization and New Paradigm for Intelligent Manufacturing from the Ministry of Industry and Information Technology of China in 2016 and in part by the Fundamental Research Funds for the Central Universities of Central South University (2021zzts0700).

Institutional Review Board Statement: Not applicable.

Informed Consent Statement: Not applicable.

Data Availability Statement: The data presented in this study are available upon request from the first author. The data are not publicly available due to intellectual property protection.

Conflicts of Interest: The authors declare no conflict of interest.

References

1. Wachtmeister, M. Overview and Analysis of Environmental and Climate Policies in China's Automotive Sector. *J. Environ. Dev.* **2013**, *22*, 284–312. [CrossRef]
2. Hu, X.; Li, S.E.; Yang, Y. Advanced Machine Learning Approach for Lithium-Ion Battery State Estimation in Electric Vehicles. *IEEE Trans. Transp. Electrif.* **2015**, *2*, 140–149. [CrossRef]
3. Xiong, R.; Zhang, Y.; He, H.; Zhou, X.; Pecht, M.G. A Double-Scale, Particle-Filtering, Energy State Prediction Algorithm for Lithium-Ion Batteries. *IEEE Trans. Ind. Electron.* **2017**, *65*, 1526–1538. [CrossRef]
4. Xiong, R.; Yu, Q.; Wang, L.Y.; Lin, C. A novel method to obtain the open circuit voltage for the state of charge of lithium ion batteries in electric vehicles by using H infinity filter. *Appl. Energy* **2017**, *207*, 346–353. [CrossRef]
5. Xiong, R.; Li, L.; Tian, J. Towards a smarter battery management system: A critical review on battery state of health monitoring methods. *J. Power Source* **2018**, *405*, 18–29. [CrossRef]
6. Wang, Y.; Zhang, C.; Chen, Z. A method for joint estimation of state-of-charge and available energy of LiFePO$_4$ batteries. *Appl. Energy* **2014**, *135*, 81–87. [CrossRef]
7. Moura, S.J.; Chaturvedi, N.A.; Krstić, M. Adaptive Partial Differential Equation Observer for Battery State-of-Charge/State-of-Health Estimation Via an Electrochemical Model. *J. Dyn. Syst. Meas. Control* **2013**, *136*, 011015. [CrossRef]
8. Ng, K.S.; Moo, C.-S.; Chen, Y.-P.; Hsieh, Y.-C. Enhanced coulomb counting method for estimating state-of-charge and state-of-health of lithium-ion batteries. *Appl. Energy* **2009**, *86*, 1506–1511. [CrossRef]
9. Javid, G.; Abdeslam, D.O.; Basset, M. Adaptive Online State of Charge Estimation of EVs Lithium-Ion Batteries with Deep Recurrent Neural Networks. *Energies* **2021**, *14*, 758. [CrossRef]
10. Liu, Y.; Li, J.; Zhang, G.; Hua, B.; Xiong, N. State of Charge Estimation of Lithium-Ion Batteries Based on Temporal Convolutional Network and Transfer Learning. *IEEE Access* **2021**, *9*, 34177–34187. [CrossRef]
11. Shen, Y. Adaptive online state-of-charge determination based on neuro-controller and neural network. *Energy Convers. Manag.* **2010**, *51*, 1093–1098. [CrossRef]
12. Chen, M.; Rincon-Mora, G.A. Accurate Electrical Battery Model Capable of Predicting Runtime and I–V Performance. *IEEE Trans. Energy Convers.* **2006**, *21*, 504–511. [CrossRef]
13. Chen, N.; Zhang, P.; Dai, J.; Gui, W. Estimating the State-of-Charge of Lithium-Ion Battery Using an H-Infinity Observer Based on Electrochemical Impedance Model. *IEEE Access* **2020**, *8*, 26872–26884. [CrossRef]
14. Liu, C.; Liu, W.; Wang, L.; Hu, G.; Ma, L.; Ren, B. A new method of modeling and state of charge estimation of the battery. *J. Power Source* **2016**, *320*, 1–12. [CrossRef]
15. Liu, Z.; Qiu, Y.; Feng, J.; Chen, S.; Yang, C. A simplified fractional order modeling and parameter identification for lithium-ion batteries. *J. Electrochem. Energy Convers. Storage* **2022**, *19*, 021001. [CrossRef]
16. Li, J.; Wang, L.; Lyu, C.; Liu, E.; Xing, Y.; Pecht, M. A parameter estimation method for a simplified electrochemical model for Li-ion batteries. *Electrochim. Acta* **2018**, *275*, 50–58. [CrossRef]
17. Hu, M.; Li, Y.; Li, S.; Fu, C.; Qin, D.; Li, Z. Lithium-ion battery modeling and parameter identification based on fractional theory. *Energy* **2018**, *165*, 153–163. [CrossRef]
18. Xiong, R.; Tian, J.; Mu, H.; Wang, C. A systematic model-based degradation behavior recognition and health monitoring method for lithium-ion batteries. *Appl. Energy* **2017**, *207*, 372–383. [CrossRef]
19. Mawonou, K.S.; Eddahech, A.; Dumur, D.; Beauvois, D.; Godoy, E. Improved state of charge estimation for Li-ion batteries using fractional order extended Kalman filter. *J. Power Source* **2019**, *435*, 226710. [CrossRef]
20. Plett, G. Extended Kalman filtering for battery management systems of LiPB-based HEV battery packs: Part 1. Background. *J. Power Source* **2004**, *134*, 252–261. [CrossRef]
21. Jiang, Z.; Shi, Q.; Wei, Y.; Wei, H.; Gao, B.; He, L. An Immune Genetic Extended Kalman Particle Filter approach on state of charge estimation for lithium-ion battery. *Energy* **2021**, *230*, 120805.
22. Xiaosong, H.; Fengchun, S.; Zou, Y. Estimation of State of Charge of a Lithium-Ion Battery Pack for Electric Vehicles Using an Adaptive Luenberger Observer. *Energies* **2010**, *3*, 1586–1603.
23. Xu, J.; Mi, C.; Cao, B.; Deng, J.; Chen, Z.; Li, S. The State of Charge Estimation of Lithium-Ion Batteries Based on a Proportional-Integral Observer. *IEEE Trans. Veh. Technol.* **2014**, *63*, 1614–1621. [CrossRef]
24. Zhang, F.; Liu, G.; Fang, L. Estimation of Battery State of Charge with H infinity Observer: Applied to a Robot for Inspecting Power Transmission Lines. *IEEE Trans. Ind. Electron.* **2012**, *59*, 1086–1095. [CrossRef]
25. Zhang, Y.; Zhang, C.; Zhang, X. State-of-charge estimation of the lithium-ion battery system with time-varying parameter for hybrid electric vehicles. *IET Control. Theory Appl.* **2014**, *8*, 160–167. [CrossRef]
26. Kim, I.-S. The novel state of charge estimation method for lithium battery using sliding mode observer. *J. Power Source* **2006**, *163*, 584–590. [CrossRef]

27. Chen, X.; Shen, W.; Dai, M.; Cao, Z.; Jin, J.; Kapoor, A. Robust Adaptive Sliding-Mode Observer Using RBF Neural Network for Lithium-Ion Battery State of Charge Estimation in Electric Vehicles. *IEEE Trans. Veh. Technol.* **2016**, *65*, 1936–1947. [CrossRef]
28. Kim, I.-S. Nonlinear State of Charge Estimator for Hybrid Electric Vehicle Battery. *IEEE Trans. Power Electron.* **2008**, *23*, 2027–2034. [CrossRef]
29. Wang, Y.; Zhang, C.; Chen, Z. A method for state-of-charge estimation of LiFePO4 batteries at dynamic currents and temperatures using particle filter. *J. Power Source* **2015**, *279*, 306–311. [CrossRef]
30. Li, J.; Barillas, J.K.; Guenther, C.; Danzer, M.A. Multicell state estimation using variation based sequential Monte Carlo filter for automotive battery packs. *J. Power Source* **2015**, *277*, 95–103. [CrossRef]
31. Charkhgard, M.; Farrokhi, M. State-of-Charge Estimation for Lithium-Ion Batteries Using Neural Networks and EKF. *IEEE Trans. Ind. Electron.* **2011**, *57*, 4178–4187. [CrossRef]
32. Chen, Z.; Fu, Y.; Mi, C. State of Charge Estimation of Lithium-Ion Batteries in Electric Drive Vehicles Using Extended Kalman Filtering. *IEEE Trans. Veh. Technol.* **2013**, *62*, 1020–1030. [CrossRef]
33. Zhang, J.; Xia, C. State-of-charge estimation of valve regulated lead acid battery based on multi-state Unscented Kalman Filter. *Int. J. Electr. Power Energy Syst.* **2011**, *33*, 472–476. [CrossRef]
34. Ouyang, Q.; Chen, J.; Wang, F.; Su, H. Nonlinear Observer Design for the State of Charge of Lithium-Ion Batteries. *IFAC Proc. Vol.* **2014**, *47*, 2794–2799. [CrossRef]
35. Chen, B.; Zhang, H.; Lin, C. Observer-Based Adaptive Neural Network Control for Nonlinear Systems in Nonstrict-Feedback Form. *IEEE Trans. Neural Netw. Learn. Syst.* **2015**, *27*, 89–98. [CrossRef]
36. Zhao, Z.; Ren, Y.; Mu, C.; Zou, T.; Hong, K.-S. Adaptive Neural-Network-Based Fault-Tolerant Control for a Flexible String with Composite Disturbance Observer and Input Constraints. *IEEE Trans. Cybern.* **2021**, 1–11. [CrossRef]
37. Chen, J.; Ouyang, Q.; Xu, C.; Su, H. Neural Network-Based State of Charge Observer Design for Lithium-Ion Batteries. *IEEE Trans. Control Syst. Technol.* **2017**, *26*, 313–320. [CrossRef]
38. Petras, I. *Fractional-Order Nonlinear Systems: Modeling, Analysis and Simulation*; Springer Science & Business Media: Berlin/Heidelberg, Germany, 2011.

Article

Experimentally Validated Coulomb Counting Method for Battery State-of-Charge Estimation under Variable Current Profiles

Bachir Zine [1], Haithem Bia [1], Amel Benmouna [2,3,*], Mohamed Becherif [3,*] and Mehroze Iqbal [3]

[1] Mechanical Engineering Department, Faculty of Technology, University of Eloued, El Oued 39000, Algeria
[2] ESTA, School of Business and Engineering, 90000 Belfort, France
[3] Femto-ST, FCLAB, University Bourgogne Franche-Comte, CNRS, UTBM, 91110 Belfort, France
* Correspondence: amel.benmouna@utbm.fr (A.B.); mohamed.becherif@utbm.fr (M.B.)

Abstract: Battery state of charge as an effective operational indicator is expected to play a crucial role in the advancement of electric vehicles, improving the battery capacity and energy utilization, avoiding battery overcharging and over-discharging, extending the battery's useful lifespan, and extending the autonomy of electric vehicles. In context, this article presents a computationally efficient battery state-of-charge estimator based on the Coulomb counting technique with constant and variable discharging current profiles for an actual battery pack in real time. A dedicated experimental bench is developed for validation purposes, where pivotal measurements such as current, voltage, and temperature are initially measured during the charging/discharging cycle. The state of charge thus obtained via these measurements is then compared with the value estimated through the battery generic model. Detailed analysis with conclusive outcomes is finally presented to exhibit the flexible nature of the proposed method in terms of the precise state-of-charge estimation for a variety of batteries, ranging from lead–acid batteries for domestic applications to Li-ion batteries inside electric vehicles.

Keywords: state of charge; electric vehicle; coulomb counting approach; battery generic model

Citation: Zine, B.; Bia, H.; Benmouna, A.; Becherif, M.; Iqbal, M. Experimentally Validated Coulomb Counting Method for Battery State-of-Charge Estimation under Variable Current Profiles. *Energies* **2022**, *15*, 8172. https://doi.org/10.3390/en15218172

Academic Editor: Siamak Farhad

Received: 28 September 2022
Accepted: 25 October 2022
Published: 2 November 2022

Publisher's Note: MDPI stays neutral with regard to jurisdictional claims in published maps and institutional affiliations.

1. Introduction

The battery state of charge (SOC) for electric vehicles is equivalent to the oil meter for conventional fuel vehicles. Typically, the relation between the electrochemical reactions and SOC is complex and difficult to determine. In addition, for vehicles, the working conditions are challenging and complex. It is therefore very difficult to obtain precise SOC as it is a hidden state function of electrochemical reactions inside the battery. It is possible to separate the battery SOC estimation strategies into four classes [1]: characteristic param-driven approach [2–7], an integral estimation method [8–11], a physical model-driven method [12–16], and a data-driven approach [17–25], as illustrated in Figure 1.

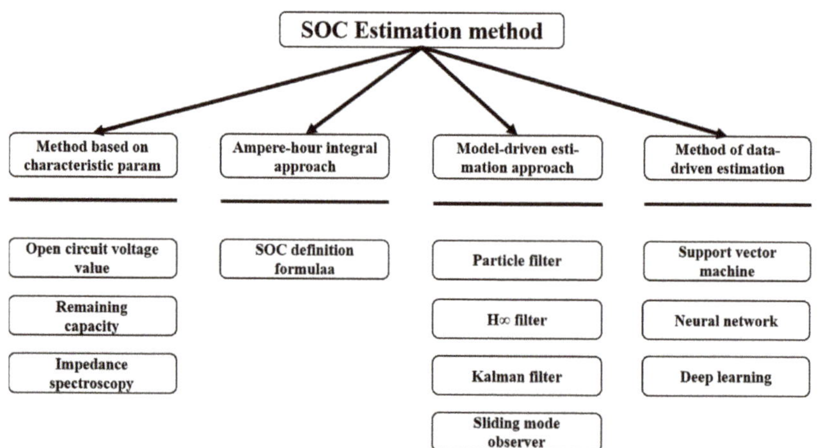

Figure 1. Classification of SOC estimation methods [26].

Subsequently, Table 1 sums up the benefits and drawbacks of four SOC assessment strategies, which are presented in Figure 1. Due to the importance of this topic, SOC estimation is rigorously investigated by several researchers. The work of [27] evaluated the SOC estimation via two model-based techniques that are the extended Kalman filter and the adaptive dual extended Kalman filter influenced by a fuzzy inference system. The experimental results demonstrate that the last technique provides a more precise indication of the battery SOC. The reference [28] is focused on optimizing the battery state of health within the domain of satellite applications. To enhance the performance of the system, the authors have taken into consideration different constraints linked to the battery SOC. For this, the Coulomb counting technique is used to estimate the battery SOC. Another study [29] treated the battery SOC as an operational indicator, which can influence the energy consumption of the electric vehicle [30]. Therefore, the SOC estimator is defined as the crude power consumption of the studied electrical vehicle (for more details, the reader can consult [31,32]).

The goal of this work is to estimate the SOC of a domestic-grade lead–acid battery through empirical measurements of battery voltage and current using the Ampere-hour integral estimation method with constant and variable C-rates for discharge current. There are many studies on the SOC estimation of lead–acid and Li-ion batteries, especially the Coulomb counting method [33,34]. However, the research work conducted in this article differentiates by investigating the discharging behavior of a lead–acid battery in extended and varying conditions.

Table 1. SOC estimation methods: benefits, drawbacks, accuracy, and robustness. Reprinted from Refs. [35,36].

Methodology	Benefits	Drawbacks	Precision	Sturdiness
Method based on characteristic param	- simpler implementation - lower computing burden - Real-time application	- easily influenced by factors of uncertainty - standard OCV or information calibration is required	poor	good

Table 1. *Cont.*

Methodology	Benefits	Drawbacks	Precision	Sturdiness
Ampere-hour integral approach	- simpler implementation - lower computational burden - Real-time application	- exact initial value of SOC is required - use of a high precision sensor causes a cumulative error. - is affected by drift noise and ageing.	average	poor
Model-driven estimation approach	- high precision - closed loop regulation - Real-time application - Adaptive	- requires model precision - computational complexity - divergence of predicted outcomes	excellent	excellent
Method of data-driven estimation	- high precision - Suitable for nonlinear implementation	- computational complexity - offline training	excellent	poor

For that, a dedicated test bench is prepared, and cross-validation is conducted under real-time scenarios and simulations. Finally, the performance of the proposed method is depicted by an evident comparison with the data provided by the manufacturer under variable operating conditions and with the generic battery model. It is worth mentioning here that the proposed method is generic and flexible in terms of its application, which is especially true for domestic-grade batteries as well as for modern Li-ion batteries present in electric vehicles. Moreover, the hysteresis effect is also taken into account considering the studied lead–acid battery.

The remainder of the paper is presented as: In Section 2, the battery management system is summarized by citing some references. The Coulomb counting method is highlighted in Section 3. In Section 4, the generic battery model is developed and comprehensively discussed. The experimental setup, which exhibits the feasibility and the applicability of the proposed approach for the lead–acid battery to attain adequate SOC values, is highlighted in Section 5. Section 6 is dedicated to the obtained results that are discussed in detail. Finally, the conclusions and the relevant perspectives are drawn in Section 7.

2. Battery Management System

The battery management system (BMS) is a technology that performs the micro-management of the battery pack in terms of its state of health (SOH). Figure 2 depicts a general layout of a typical BMS, which constitutes numerous sensors, actuators, controllers, connection lines, etc. The bidirectional communication between the control unit of BMS and external bodies such as the human media interface is typically carried out via CAN (the communication bus controller area network). Besides maintaining reliability during normal and abnormal operating conditions, adequate provision of the battery state of charge to the vehicle's vehicle control unit (VCU) is among the vital tasks of BMS.

A well-built BMS aids in collecting critical data in real time, such as electrical measurements, temperature, and other relevant data via inherent sampling hardware. The collected data is then exploited via embedded algorithms and strategies to estimate the battery states, such as SOC [37], SOH [38], SOP [39], and RUL [40]. These states are ultimately fed to VCU, in order to facilitate effective power management.

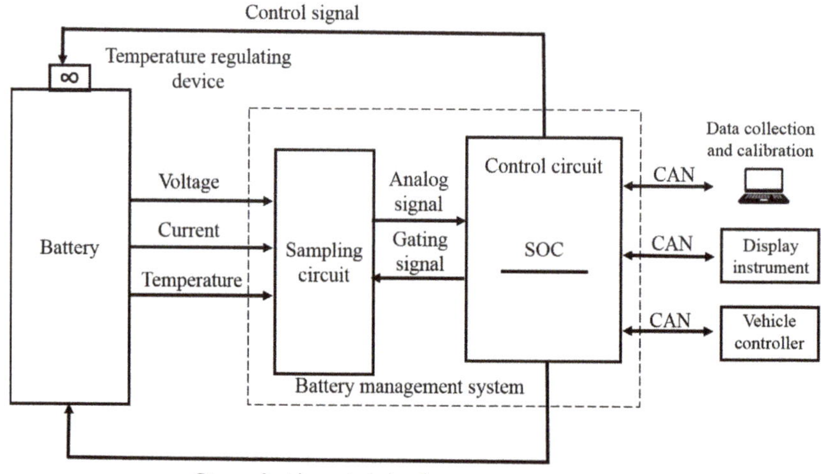

Figure 2. Schematic diagram of a typical BMS [26].

3. Coulomb Counting Method

The Coulomb counting method (known as the current integration method) is chosen as the baseline in this article and is among the most exploited method [41]. It consists of measuring the battery open circuit voltage at the start-up to estimate the initial SOC using the battery datasheet information. Then, the battery current information is integrated to estimate the amount of charge delivered (or recovered) by the batteries. The overall concept is presented in Equation (1), where knowledge of the initial SOC is detrimental to determining the state of charge [24,42,43]. Based on this approach, the SOC for the battery pack is calculated as follows [44]:

$$ \text{SOC} = \text{SOC}_0 - \frac{1}{C_N} \int_{t0}^{t} \eta \cdot I(\tau) d\tau \tag{1} $$

If the initial value of the charge state SOC_0 is specified, or imposed (technically, most researchers impose the initial state as to be fully charged or fully discharged), the Coulomb counter provides precise estimation with relative ease and simplicity [45]. On the contrary, this method is less accurate if the SOC_0 is unknown.

The Coulomb counting strategy computes the remaining stored energy essentially by collecting the charge moved in or out of the battery. The precision of this strategy depends fundamentally on a real estimation of the battery current and a precise assessment of the initial SOC. With pre-knowledge of the initial SOC, which also can be stored at the end of the vehicle trip in a flash memory to be reused as the initial SOC for the next trip (and neglecting the battery self-discharge), the battery SOC can be determined by computing the stored and the released energy flows over the operating time. Nevertheless, the stored energy in the battery is not completely available to be used due to the DOD (depth of discharge) which is a quantity of energy to keep inside the battery to avoid permanent damage. Moreover, there are losses during the charging and discharging process. For an exact SOC assessment, this effect should be considered [46]. Futhermore, the SOC must be recalibrated consistently, and the discharge limit should be considered for an exact assessment.

4. Generic Battery Model

The battery model can be obtained by considering a voltage source in series with constant resistance, as appears in Equation (2) [11].

$$V_{batt} = E_0 - K\frac{Q}{Q - i_t}\cdot i_t - R\cdot I + A\cdot \exp\left(-B\cdot i_t\right)K\frac{Q}{Q - i_t}\cdot i^* \tag{2}$$

As depicted in Figure 3, the special feature of this model is the use of filtered current (i^*) through the polarization resistance. This filtered current solves the problem of the algebraic loop due to the simulation of power systems in Simulink [47]. Despite the existence of hysteresis between the charging and discharging of the battery voltage, this model is still valid for both charge and discharge cycles [48]. The battery models can be obtained using:

✓ The charge model ($i^* < 0$)

$$f_1(it, i^*, exp, batt\ type) \ = \ E_0 \ - K\frac{Q}{i_t - 0.1Q}\cdot i* - K\ \frac{Q}{i_t - Q}\cdot i_t + \mathrm{Exp}\ (t) \tag{3}$$

✓ The discharge model ($i^* > 0$)

$$f_1(it, i^*, exp, batt\ type) \ = \ E_0 \ - K\frac{Q}{i_t - Q}\cdot i* - K\ \frac{Q}{i_t - Q}\cdot i_t + \mathrm{Exp}\ (t) \tag{4}$$

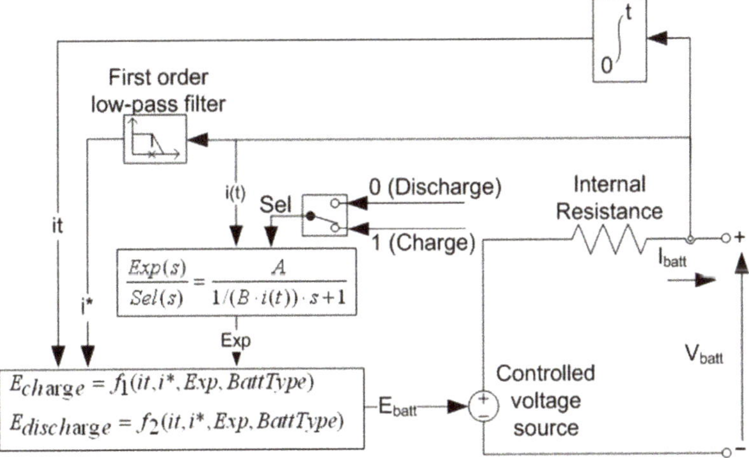

Figure 3. Graphical description of the employed battery model.

The employed model is simplistic and cannot exactly mimic the complex electro-chemical reactions taking place within the battery under the influence of actual operating conditions. However, this model is still adequate from the computational side, and can suffix for a range of applications. Especially, for the discharge state of lead acid batteries with different currents, which is extensively studied, where the discharge curve of the simulations matches with the discharge curve of the experimental work and with that of the datasheet. Therefore, even with assumptions and limitations, the utilized model suffixes the fundamental needs of this research work. It is worth mentioning here that the hysteresis phenomenon for the lead–acid battery is considered here. The effect can be seen in Figure 4, exhibiting that the exponential voltage increases when the battery is charging, while during the discharging, the exponential voltage decreases immediately.

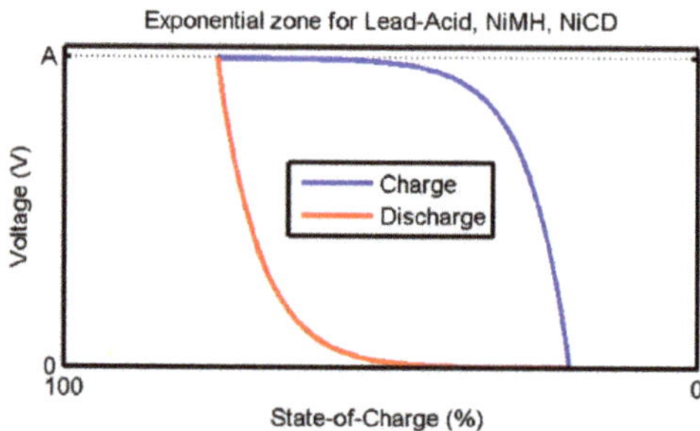

Figure 4. Hysteresis effect associated with the lead–acid battery.

Application: From the discharge curve provided by the manufacturer, the extracted parameters are presented in Table 2. The model also has its assumptions and limitations as follows:

✓ Limitations

- The minimum no-load battery voltage is 0 V and the maximum battery voltage is equal to $2 \times E_0$.
- The minimum capacity of the battery is 0 Ah and the maximum capacity is Q_{max}.

✓ Assumptions

- The internal resistance is supposed as a constant value during the charge and the discharge cycles and does not vary with the amplitude of the current.
- The parameters of the model are deduced from discharge characteristics and assumed to be the same for charging.
- The capacity of the battery does not change with the amplitude of the current (no Peukert effect).
- The model does not take the temperature into account.
- The self-discharge of the battery is not represented. It can be represented by adding a large resistance in parallel with the battery terminals.
- The battery has no memory effect.

Table 2. Battery parameters: discharge at variable C-rate.

C-Rates of Discharge	0.25 C	0.17 C	0.09 C	Different C-Rates
Nominal voltage (V)	12	12	12	12
Rated capacity (Ah)	52	52	52	52
Initial state-of-charge (%)	100	100	100	100
Max. capacity (Ah)	40.1999	53.8	47.5	47.5
Fully charged voltage (V)	12.7	12.8	12.9	12.9
Nominal discharge current (A)	13	8.84	4.68	12
Internal resistance (Ω)	0.0055	0.0055	0.0055	0.0055
Capacity (Ah) @ nominal voltage	38.2	44.80	46	46
Exp. zone [Voltage(V), capacity(Ah)]	[12.7, 2.5]	[12.8, 1.9]	[12.9, 2.225]	[12.9, 2.225]

5. Experimental Method and Description

Battery specification: The battery employed in this research is a valve-regulated lead–acid (VRLA) type with a nominal voltage of 12 V and a nominal capacity of 52 Ah. The recommended voltage when charged for standby use is 1.75 V for a single cell (6 cells in series). Further features of this battery are presented in Table 3.

Table 3. Lead–acid battery parameters.

Cells per unit	6
Voltage per unit	12 V
Capacity	52 Ah @ 20 h-rate to 1.75 V per cell @25 °C (77 °F)
Weight	Approx. 18 Kg (39.68 lbs.)
Maximum discharge current	500 A (5 s)
Internal resistance	Approx. 5.5 mΩ
Operating temperature range	Discharge: −15 °C~50 °C (5 °F~122 °F) Charge: −15 °C~40 °C (5 °F~104 °F) Storage: −15 °C~40 °C (5 °F~104 °F)
Float charging voltage	13.5 to 13.8 VDC/unit Average at 25 °C (77 °F)
Maximum charging current limit	15.6 A
Equalization and cycle service	14.4 to 15.0 VDC/unit Average at 25 °C (77 °F)
Self discharge	Batteries can be stored for 6 months at 25 °C (77 °F).

Measurement test bench: The measurement test bench is presented in Figure 5, which further consists of four interconnected components.

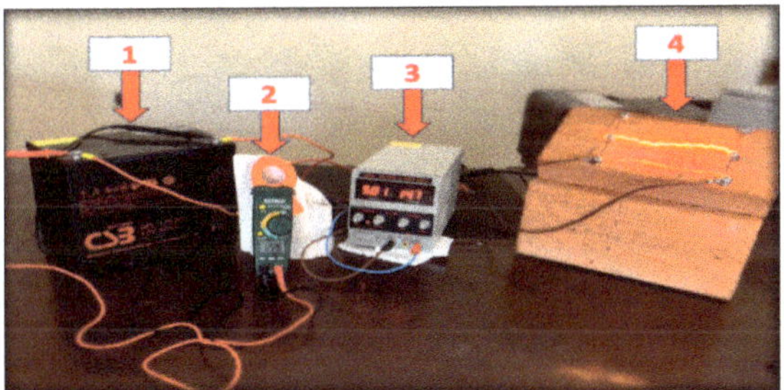

Figure 5. The test bench.

1. Lead–acid battery
2. Clamp meter
3. DC power supply
4. Variable resistor (handmade)

Computing parameters: The proposed computing method is implemented in Matlab/Simulink version 2015b and has the following attributes and parameters: variable step, ode45 (Dormand–Prince), relative tolerance = 1×10^{-3}, time tolerance = $10 \times 128 \times$ eps, zero crossing control = use local settings, algorithm = non adaptive, number of consecutive zero crossings = 1000, number of consecutive min steps = 1.

Charging process test: The battery is charged via an external DC power supply, and the initial voltage at the time of testing is effectively set at 14.7 V. The current maximum limit is set to 5.03 A (this is the maximum available value). The battery is connected to the power supply and is left until its fully charged. The voltage and current are measured using a voltmeter and clamp meter every 10 min. The end-of-charge voltage equals $V_{finish\text{-}ocv}$ = 13.96 V and the end-of-charge current (the minimum value of the current in the vicinity of full charge) is I_{min} = 0.29 A. When the charging current reaches the minimum value of 0.29 A (indicated by the manufacturer), the battery is disconnected, and the DC power voltage supply is turned off. This test was spanned over 02 days, the duration of this test, therefore, equals T_{ch} = 1440 min or T_{ch} = 24 h.

Discharging process test: Two experiments are conducted in total with relevance to the discharging test:

- *Discharging with constant C-rates*: a variable resistor is connected (for adjustment to variable C-rates, that is to reach discharge current at 0.25 C, 0.17 C, and 0.09 C respectively) with the battery. Consequently, the voltage and the discharge current are measured after every 5 min by using a clamp meter. After 3.05, 5.41, and 10.25 h, respectively (corresponding to the associated C-rate), it reaches the lower permissible limit of discharge voltage (cut-off voltage) 10.10, 10.37, and 10.39 V each. Afterwards, the discharge resistance is disconnected. The discharge resistance equaled R_{disch} = [0.96–1] Ω, [1.42–1.50] Ω, and [2.69–3] Ω, with the ambient temperature T = 26.7°, 28.6°, and 26.1 °C.

- *Discharging with variable C-rates*: the variable resistor is connected (for adjustment to variable C-rates discharge current 0.23 C, 0.115 C, and 0.057 C, respectively. These values of C-rates correspond to having three resistances connected in parallel. The value of each resistance is 3.225 Ω. During the first period, the discharge is performed at 0.23 C until 1.11 h. After that, the first resistance is disconnected, then the discharge continues at 0.115 C until 4.44 h. At this point, the second resistance is also disconnected. The discharge is pursued then with 0.057 C until 7.23 h, and at this moment, even the last resistance is disconnected, and the experiment is stopped. The battery voltage and discharging current are measured every 5 min with a clamp meter during the whole experimental session. After 7.23 h, the end of discharge is reached (cut-off voltage) which is equal to 10.39 V. At this point, the discharge resistance is completely disconnected. It is also worth mentioning that the ambient temperature equaled T = 27.1 °C (approx.) during the experimental session.

6. Results and Discussion

As depicted in Figure 6: the discharge voltage at 0.25 C, 0.17 C, and 0.09 C decreases to 10.10 V, 10.37 V, and 10.39 V, respectively, during 3.05, 5.41, and 10.25 h. These values are called the cut-off voltages, i.e., when the battery discharge voltage has reached the lowest permissible value. At this point, the battery is disconnected to avoid permanent damage. The 10.10 V, 10.37 V, and 10.39 V values are not indicative of the total discharge (SOC = 0%) but correspond to the minimum SOC value in the vicinity of 20%, also known as depth of discharge (DoD) [49].

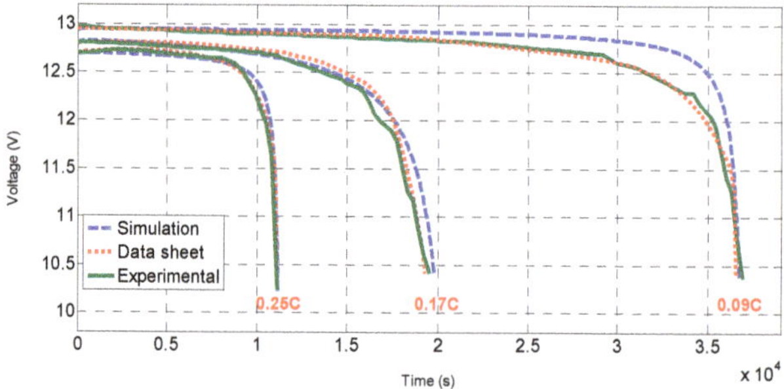

Figure 6. The discharge voltage vs. time at 0.25/0.17 and 0.09 C.

It can be observed that the experimental discharge curve is very close to the simulated one, except for the sensor's noise and deviation in the collected points. This difference shows that battery power varies according to the operating conditions and user parameters (charging mode, ambient temperature, discharge current, etc.). It should be noted that the manufacturer-provided data normally correspond to ideal conditions (T = 25 °C, Idisch = 13 A, 8.84 A, and 4.68 A, respectively) and the battery's aging status (SOH = 100%). which means a new battery. So, during the experiments, it is found that the 03 sets (datasheet, simulated, and experimental) are close, this indicates that the battery model used shows a very high precision in the discharge phase of the battery.

Figure 7 depicts the results corresponding to estimated SOC evolution with discharging at fixed C rates. It can be observed that the battery is initially fully charged (SOC_0 = 100%) and that the battery charging status decreases to a minimum value (SOC_{min} = 20%) after 3.05, 5.41, and 10.25 h, respectively. This value must not exceed to prevent the battery from being permanently damaged, as depicted in [50] for the application of hybrid electric vehicles. The straight line is the SOC curve calculated by the Coulomb counting estimator, i.e., the battery is discharged at a constant discharge current (I_{disch} = 13 A, 8.84 A, and 4.68 A, respectively). This approach facilitates the battery's SOC evolution over the entire discharge cycle (3.05 h, 5.41 h, and 10.25 h, respectively).

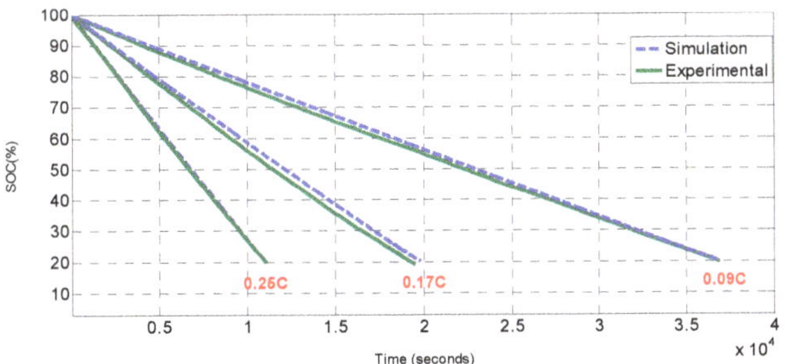

Figure 7. The SOC as a function of time at fixed C rates.

All the tests evidently reflect that the experimental SOC and the simulated ones are very close. This shows that this technique is very accurate and useful thanks to its simplicity of implementation and calculations. Additionally, it is easy to implement, since it only exploits the data received from current and voltage sensors.

Figure 8 exhibits the result of battery discharge at variable C rates using a handmade variable resistor (three equal resistors connected in parallel), where the total equivalent resistor equals R_{eq1} = 1.075 ohms. At the beginning, all resistors are connected and a discharging rate of 0.23 C is applied from [0 to 1.11 h]. Then, one resistance is disconnected from the total equivalent resistors to obtain R_{eq2} = 2.15 ohms, corresponding to a discharge rate of 0.115 C from [1.11 to 4.44 h]. Finally, a second resistance is disconnected, which means that only one resistance is kept connected, with a value of 3.225 ohms, corresponding to a discharging rate of 0.057 C from [4.44 to 7.23 h].

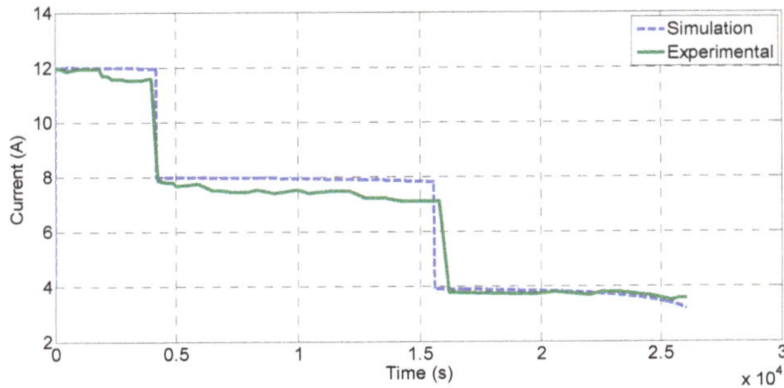

Figure 8. The discharge current as a function of time at variable C-rates.

Therefore, it is noticed that the current is constant during each C-rates of the discharging range. This is a safe and effective method. Where the minimum value of discharging current is 2.90 A, which corresponds to a cutting voltage of 10.39 V.

Figure 9 illustrates that with the applied variable C-rates, the discharge voltage decreases to 10.39 V over the period of 7.23 h. From the presented curve, it can be noticed that the three ranges express the discharging process with the different C-rates: 0.23 C,

0.115 C, and 0.057 C, respectively. The two curves (simulated and experimental) are almost identical, exhibiting that the generic battery presents the real behavior of the battery pack during the discharging process with very high precision.

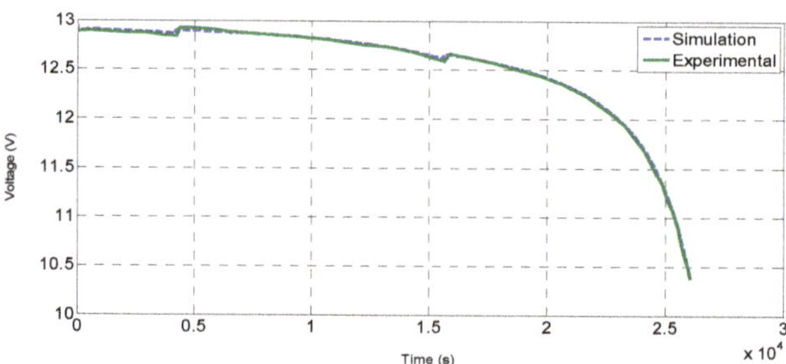

Figure 9. The discharge voltage as a function of time at variable C-rates.

The evolution of estimated SOC under the influence of variable C rates is depicted in Figure 10. It can be observed that the battery is initially fully charged (SOC0 = 100%) and the battery charging status decreases after 7.23 h to its minimum threshold (SOC = 20%), corresponding to the recommended value of the lead–acid battery (SOC$_{min}$ = 20%). Again, this value must not exceed to prevent the battery from being permanently damaged. Furthermore, the theoretical estimation matches the experimental findings.

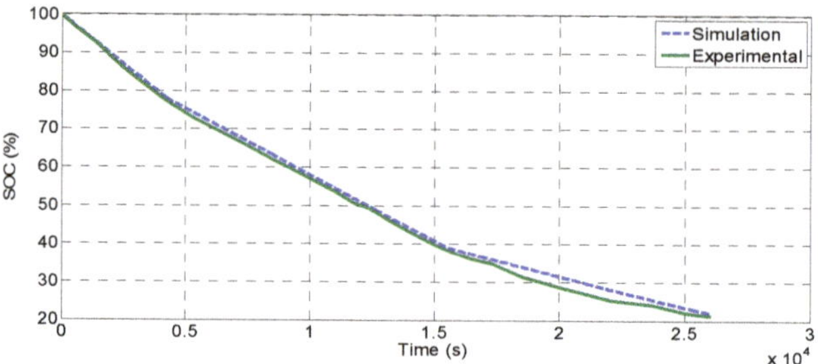

Figure 10. The SOC as a function of time at variable C rates.

The SOC curve calculated by the Coulomb counting estimator is nearly a straight line (in their specific regions), i.e., the discharging test is performed individually under variable discharge currents for sustained periods (I_{disch} = 12 A, 6 A then 3 A, respectively). It can be observed that the SOC$_{experimental}$ and SOC$_{simulated}$ are almost the same, which shows that this technique is accurate and useful.

Finally, Figure 11 exhibits the two curves (datasheets and simulation), depicting that the initial value of the battery charge current is equal to 5.5 A and 5.03 A. It can be noticed that the charging current stays constant for 3.50 h at a value of 5.5 A in datasheet/simulations, while 5.03 A in the experimental curve. Then the value decreases until 0.29 A in all the cases whether experimental or simulations. One can distinguish two areas on the curve where the first is from [0 to 3.20] h in the experimental curve and from [0 to 3.50] h in the simulation and datasheet curves. These curves are the representation of the charging process via constant current (CC) and by a constant voltage (CV). The only difference between the three curves (datasheet, simulation, and experimental) is in the zone of charge by a constant current and exactly in the initial value given, wherein the experimental test, it is set to a value of 5.03 A only (which is the maximum available power supply value). In contrast, the value of the datasheet and simulation is 5.5 A.

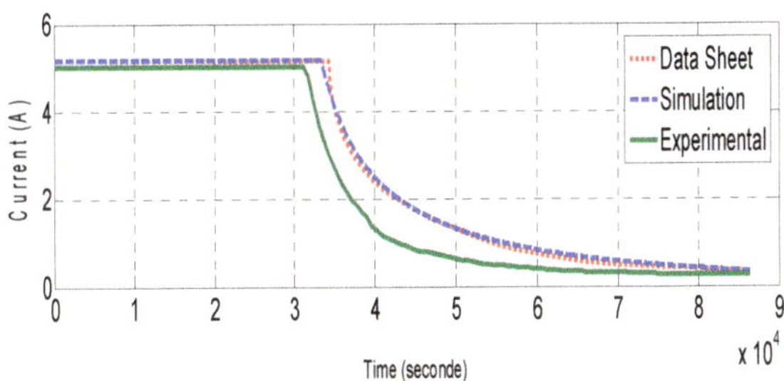

Figure 11. The charging current as the function of time.

Figure 12 indicates the fact that the battery voltage at the start of this test is set to 11.6 V in the three curves (experimental, datasheets, and simulations). Moreover, the initial charge condition is SOC_0 = 20% [44]. It can be observed that the charge voltage rises rapidly up to 14.7 V and stabilizes at this value, i.e., two curve areas are distinguished where the first is from [0 to 3] h in the experimental/datasheet curve, and from [0 to 2.5] h in the simulated curve, intended for the constant current (CC) charging mode [45]. The second distinguishable area is from [3 to 20] h intended for the charging mode using a constant voltage (CV) mode.

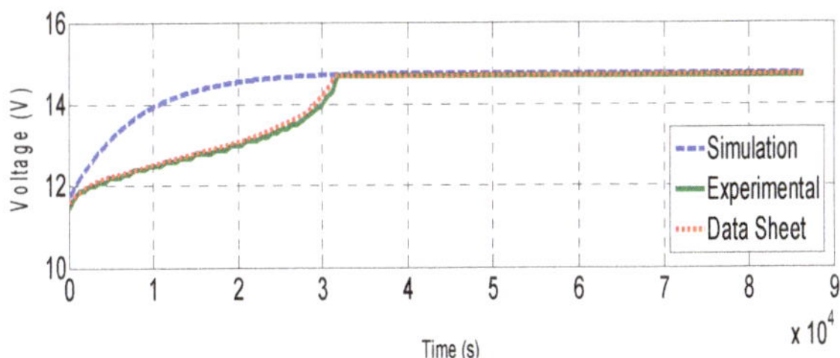

Figure 12. Charging voltage as the function of time.

The only difference between the three curves (experimental, datasheet, and simulated) is in the zone of charge by a constant current (the duration of this phase is different as in the datasheet/experimental is 3 h and in simulations is 2.5 h). This is because the simulation parameters in the discharge phase are the same as for the charging phase. After all, the charging phase is dependent on numerous factors such as charging mode, initial SoC, ambient temperature, and mode of charging. This leads to a charge period of 24 h when the end-of-charge current is equal to 0.29 A to prevent overcharging the battery by turning off the DC power supply.

In Figure 13, the SOC_0 equals 20% at time t = 0 and increases with the relation of charging up to 100%, i.e., the battery is fully charged.

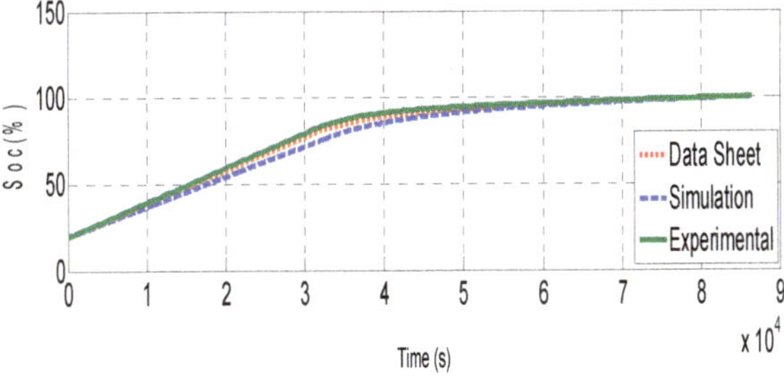

Figure 13. The SOC as function of time.

There are two parts in the curve, from [0 to 3.5] h is the first portion, which corresponds to the constant current charging mode. It has almost a linear shape. This part is characterized by the rapid variation of the SOC as a time-dependent variable for instance at 3.5 h SOC (experimental) = 85%, SOC (datasheet) = 83% and SOC (simulation) = 80%. The second part is in the form of a nonlinear curve line. It corresponds to the charging mode at a constant voltage; it is characterized by the slow variation of the SOC. For example, during [3.5–24] h, the SOC (experimental) increases by 15%, SOC (datasheet) increases by 17%, and SOC (simulation) increased by 20%.

Hence, it can be said that the battery charge capacity is dependent on the charging mode, it is faster in CC mode and slower in CV mode, which is why the new charger is faster to gain more time, especially in the lead–acid batteries.

From Figure 13, the Coulomb counter approach provides a fair depiction of SOC during the entire charge; as a result, one can say that this method is independent of the battery's model and technology.

7. Conclusions and Perspectives

This study presents a Coulomb counting technique-based battery state-of-charge estimator with constant and variable discharging current profiles for a real battery pack in a real-time environment. The following conclusions are drawn from this study:

In the charging mode, the battery model used in the MATLAB Simulink is less efficient. Regardless, it is widely used and has excellent battery economy when discharged at a constant current. That is the reason why it is also included for the sake of comparison.

The only short-coming of the Coulomb counting method is the difficulty of estimating the initial SOC. High-precision estimation sensors (for voltage and current) are also needed, so these instruments need to be periodically modified.

Lead–acid batteries are only suitable for short-range vehicles. They remain the cheapest form of battery and are likely to be used for these purposes. A lot of useful and small-scale EVs that do not require a long-range can be made via lead–acid batteries.

Most commercial-scale EVs require an extended traveling range; therefore, modern Li-ion batteries can serve the purpose. The proposed SOC estimator is expected to provide quick and reliable information, which can be then integrated into energy management.

It is concluded that constant current discharge is quicker, which is the approach used for fast charging, and it is shown in this work that the efficiency of the generic battery model is competitive, as its findings are close to experimental work. The Coulomb counter approach is useful for estimating the battery SOC; also, in this case, the discharge with variable discharge currents is always correct.

By using simple calculations and hardware requirements, the proposed method can therefore be systematically implemented in any portable devices as well as electric cars. In this research, the ampere-hour integral method is validated experimentally via a domestic lead–acid battery. It is worth mentioning here that the proposed method is generic in its implementation. Given that, the implementation and application of modern Li-ion batteries are reserved for future studies.

Author Contributions: Conceptualization, results analysis, managerial insights, and methodology are performed by B.Z.; H.B. was involved in writing original draft preparation. Formal analysis and investigations were performed by A.B.; M.B. was involved in writing the reviews and managing the references. Finally, M.I. performed language correction and general organization. All authors have read and agreed to the published version of the manuscript.

Funding: This work was not funded by any organization and/or society in any capacity.

Conflicts of Interest: The authors have no conflict of interest to declare that are relevant to the content of this article.

Abbreviations

SOC	State of charge
Ah	Ampere hour
OCV	Open circuit voltage
AC	Alternative current
SOH	State of health
CC	Constant current
CV	Constant voltage
C_N	Nominal capacity of the battery
η	Coulomb efficiency
$I(\tau)$	The current versus time (negative during charge and positive during discharge)
V_{batt}	The battery voltage (V)
E_0	The constant battery voltage (V)
K	The polarization constant (V/(Ah)) or polarization resistance (Ω)
Q	The battery capacity (Ah)
i_t	$=\int I dt$: actual battery charge (Ah)
A	The exponential voltage (V)
B	The exponential capacity (Ah) -1
R	The internal resistance (Ω)
I	The battery current (A)
i^*	The filtered current (A)
I_{dis}	The value of the current of discharge
SOC_0	Initial state of charge
SOC_{min}	Minimum state of charge
SOC_{exp}	Experimental state of charge
SOC_{th}	Theoretical state of charge
Vch_{int}	Initial charge voltage
Vch_{end}	End charge voltage
Tch_{sum}	Simulation charge time
$Vdis_{int}$	Initial discharge voltage
$Vdis_{end}$	End discharge voltage
$Tdis_{sum}$	Simulation discharge time
VRLA	Valve regulated lead–acid battery
BMS	Battery management system
VCU	Vehicle control unit
SOP	State of power
RUL	Remaining useful life
DOD	Depth of discharge

References

1. Yang, R.; Xiong, R.; He, H.; Chen, Z. A fractional-order model-based battery external short circuit fault diagnosis approach for all-climate electric vehicles application. *J. Clean. Prod.* **2018**, *187*, 950–959. [CrossRef]
2. Rodrigues, S.; Munichandraiah, N.; Shukla, A. A review of state-of-charge indication of batteries by means of a.c. impedance measurements. *J. Power Sources* **2000**, *87*, 12–20. [CrossRef]
3. Huet, F. A review of impedance measurements for determination of the state-of-charge or state-of-health of secondary batteries. *J. Power Sources* **1998**, *70*, 59–69. [CrossRef]
4. Weng, C.; Sun, J.; Peng, H. A unified open-circuit-voltage model of lithium-ion batteries for state-of-charge estimation and state-of-health monitoring. *J. Power Sources* **2014**, *258*, 228–237. [CrossRef]
5. Dubarry, M.; Svoboda, V.; Hwu, R.; Liaw, B.Y. Capacity loss in rechargeable lithium cells during cycle life testing: The importance of determining state-of-charge. *J. Power Sources* **2007**, *174*, 1121–1125. [CrossRef]
6. Schmidt, J.P.; Chrobak, T.; Ender, M.; Illig, J.; Klotz, D.; Ivers-Tiffée, E. Studies on LiFePO$_4$ as cathode material using impedance spectroscopy. *J. Power Sources* **2011**, *196*, 5349–5355. [CrossRef]
7. He, H.; Xiong, R.; Guo, H. Online estimation of model parameters and state-of-charge of LiFePO$_4$ batteries in electric vehicles. *Appl. Energy* **2012**, *89*, 413–420. [CrossRef]
8. Zhang, Y.; Song, W.; Lin, S.; Feng, Z. A novel model of the initial state of charge estimation for LiFePO$_4$ batteries. *J. Power Sources* **2014**, *248*, 1028–1033. [CrossRef]

9. Wang, J.; Cao, B.; Chen, Q.; Wang, F. Combined state of charge estimator for electric vehicle battery pack. *Control Eng. Pract.* **2007**, *15*, 1569–1576. [CrossRef]

10. Becherif, M.; Pera, M.; Hissel, D.; Jemei, S. Estimation of the lead-acid battery initial state of charge with experimental validation. In Proceedings of the 2012 IEEE Vehicle Power and Propulsion Conference, Seoul, Korea, 9–12 October 2012; pp. 469–473.

11. Ng, K.S.; Moo, C.-S.; Chen, Y.-P.; Hsieh, Y.-C. Enhanced coulomb counting method for estimating state-of-charge and state-of-health of lithium-ion batteries. *Appl. Energy* **2009**, *86*, 1506–1511. [CrossRef]

12. Chan, C.; Lo, E.; Weixiang, S. The available capacity computation model based on artificial neural network for lead–acid batteries in electric vehicles. *J. Power Sources* **2000**, *87*, 201–204. [CrossRef]

13. Gu, W.B.; Wang, C.Y. Thermal-Electrochemical Modeling of Battery Systems. *J. Electrochem. Soc.* **2000**, *147*, 2910–2922. [CrossRef]

14. Zhang, Y.; Xiong, R.; He, H.; Shen, W. Lithium-Ion Battery Pack State of Charge and State of Energy Estimation Algorithms Using a Hardware-in-the-Loop Validation. *IEEE Trans. Power Electron.* **2016**, *32*, 4421–4431. [CrossRef]

15. Xiong, R.; He, H.; Sun, F.; Zhao, K. Evaluation on State of Charge Estimation of Batteries With Adaptive Extended Kalman Filter by Experiment Approach. *IEEE Trans. Veh. Technol.* **2012**, *62*, 108–117. [CrossRef]

16. Plett, G.L. Extended Kalman filtering for battery management systems of LiPB-based HEV battery packs Part 2. Modeling and identification. *J. Power Sources* **2004**, *134*, 262–276. [CrossRef]

17. Cuadras, A.; Kanoun, O. Soc li-ion battery monitoring with impedance spectroscopy. In Proceedings of the 2009 6th International Multi-Conference on Systems, Signals and Devices, SSD, Djerba, Taiwan, 23–26 March 2009.

18. Chau, K.; Wu, K.; Chan, C. A new battery capacity indicator for lithium-ion battery powered electric vehicles using adaptive neuro-fuzzy inference system. *Energy Convers. Manag.* **2004**, *45*, 1681–1692. [CrossRef]

19. Wang, Z.; Xu, J.; Wang, T. The online monitoring system software design and the SOC estimation algorithm research for power battery. In Proceedings of the 2013 IEEE International Conference on Vehicular Electronics and Safety, ICVES 2013, Dongguan, China, 28–30 July 2013; pp. 89–92.

20. Lv, J.; Yuan, H.; Lv, Y. Battery state-of-charge estimation based on fuzzy neural network and improved particle swarm optimization algorithm. In Proceedings of the 2012 2nd International Conference on Instrumentation and Measurement, Computer, Communication and Control, Washington, DC, USA, 8–10 December 2012; pp. 22–27.

21. Chemali, E.; Kollmeyer, P.J.; Preindl, M.; Ahmed, R.; Emadi, A.; Kollmeyer, P. Long Short-Term Memory Networks for Accurate State-of-Charge Estimation of Li-ion Batteries. *IEEE Trans. Ind. Electron.* **2018**, *65*, 6730–6739. [CrossRef]

22. He, W.; Williard, N.; Chen, C.; Pecht, M. State of charge estimation for Li-ion batteries using neural network modeling and unscented Kalman filter-based error cancellation. *Int. J. Electr. Power Energy Syst.* **2014**, *62*, 783–791. [CrossRef]

23. Kang, L.; Zhao, X.; Ma, J. A new neural network model for the state-of-charge estimation in the battery degradation process. *Appl. Energy* **2014**, *121*, 20–27. [CrossRef]

24. Junping, W.; Quanshi, C.; Binggang, C. Support vector machine based battery model for electric vehicles. *Energy Convers. Manag.* **2006**, *47*, 858–864. [CrossRef]

25. Li, I.-H.; Wang, W.-Y.; Su, S.-F.; Lee, Y.-S. A Merged Fuzzy Neural Network and Its Applications in Battery State-of-Charge Estimation. *IEEE Trans. Energy Convers.* **2007**, *22*, 697–708. [CrossRef]

26. Xiong, R.; Shen, W. *Advanced Battery Management Technologies for Electric Vehicles*; John Wiley & Sons: Hoboken, NJ, USA, 2019; pp. 1–297.

27. Yang, K.; Chen, Z.; He, Z.; Wang, Y.; Zhou, Z. Online estimation of state of health for the airborne Li-ion battery using adaptive DEKF-based fuzzy inference system. *Soft Comput.* **2020**, *24*, 18661–18670. [CrossRef]

28. Lami, M.; Shamayleh, A.; Mukhopadhyay, S. Minimizing the state of health degradation of Li-ion batteries onboard low earth orbit satellites. *Soft Comput.* **2019**, *24*, 4131–4147. [CrossRef]

29. Modi, S.; Bhattacharya, J.; Basak, P. Convolutional neural network–bagged decision tree: A hybrid approach to reduce electric vehicle's driver's range anxiety by estimating energy consumption in real-time. *Soft Comput.* **2020**, *25*, 2399–2416. [CrossRef]

30. Iqbal, M.; Becherif, M.; Ramadan, H.S.; Badji, A. Dual-layer approach for systematic sizing and online energy management of fuel cell hybrid vehicles. *Appl. Energy* **2021**, *300*, 117345. [CrossRef]

31. Kularatna, N. *Dynamics, Models, and Management of Rechargeable Batteries*; Academic Press: Cambridge, MA, USA, 2015.

32. Iqbal, M.; Laurent, J.; Benmouna, A.; Becherif, M.; Ramadan, H.S.; Claude, F. Ageing-aware load following control for composite-cost optimal energy management of fuel cell hybrid electric vehicle. *Energy* **2022**, *254*, 124233. [CrossRef]

33. Salazar, D.; Garcia, M. Estimation and Comparison of SOC in Batteries Used in Electromobility Using the Thevenin Model and Coulomb Ampere Counting. *Energies* **2022**, *15*, 7204. [CrossRef]

34. González, I.; Ramiro, A.; Calderón, M.; Calderón, A.; González, J. Estimation of the state-of-charge of gel lead-acid batteries and application to the control of a stand-alone wind-solar test-bed with hydrogen support. *Int. J. Hydrogen Energy* **2012**, *37*, 11090–11103. [CrossRef]

35. Lin, C.; Mu, H.; Xiong, R.; Shen, W. A novel multi-model probability battery state of charge estimation approach for electric vehicles using H-infinity algorithm. *Appl. Energy* **2016**, *166*, 76–83. [CrossRef]

36. Wu, G.; Lu, R.; Zhu, C.; Chan, C. An improved Ampere-hour method for battery state of charge estimation based on temperature, coulomb efficiency model and capacity loss model. In Proceedings of the 2010 IEEE Vehicle Power and Propulsion Conference, Lille, France, 1–3 September 2010; pp. 1–4.

37. Xiong, R.; Sun, F.; Chen, Z.; He, H. A data-driven multi-scale extended Kalman filtering based parameter and state estimation approach of lithium-ion polymer battery in electric vehicles. *Appl. Energy* **2014**, *113*, 463–476. [CrossRef]
38. Telmoudi, A.J.; Soltani, M.; Ben Belgacem, Y.; Chaari, A. Modeling and state of health estimation of nickel–metal hydride battery using an EPSO-based fuzzy c-regression model. *Soft Comput.* **2019**, *24*, 7265–7279. [CrossRef]
39. Sun, F.; Xiong, R.; He, H. Estimation of state-of-charge and state-of-power capability of lithium-ion battery considering varying health conditions. *J. Power Sources* **2014**, *259*, 166–176. [CrossRef]
40. Zhang, Y.Z.; Xiong, R.; He, H.W.; Pecht, M. Validation and verification of a hybrid method for remaining useful life prediction of lithium-ion batteries. *J. Clean. Prod.* **2019**, *212*, 240–249. [CrossRef]
41. Mohammed, N.; Saif, A.M. Programmable logic controller based lithium-ion battery management system for accurate state of charge estimation. *Comput. Electr. Eng.* **2021**, *93*, 107306. [CrossRef]
42. Feng, F.; Lu, R.; Zhu, C. A Combined State of Charge Estimation Method for Lithium-Ion Batteries Used in a Wide Ambient Temperature Range. *Energies* **2014**, *7*, 3004–3032. [CrossRef]
43. Zhang, S.; Guo, X.; Dou, X.; Zhang, X. A data-driven coulomb counting method for state of charge calibration and estimation of lithium-ion battery. *Sustain. Energy Technol. Assessments* **2020**, *40*, 100752. [CrossRef]
44. Gismero, A.; Schaltz, E.; Stroe, D.-I. Recursive State of Charge and State of Health Estimation Method for Lithium-Ion Batteries Based on Coulomb Counting and Open Circuit Voltage. *Energies* **2020**, *13*, 1811. [CrossRef]
45. Hinz, H. Comparison of Lithium-Ion Battery Models for Simulating Storage Systems in Distributed Power Generation. *Inventions* **2019**, *4*, 41. [CrossRef]
46. Tremblay, O.; Dessaint, L.-A.; Dekkiche, A.-I. A Generic Battery Model for the Dynamic Simulation of Hybrid Electric Vehicles. In Proceedings of the IEEE Vehicle Power and Propulsion Conference, Arlington, TX, USA, 9–12 September 2007; pp. 284–289.
47. Zine, B.; Marouani, K.; Becherif, M.; Yahmedi, S. Estimation of Battery Soc for Hybrid Electric Vehicle using Coulomb Counting Method. *Int. J. Emerg. Electr. Power Syst.* **2018**, *19*. [CrossRef]
48. Baroody, R. *UNLV Theses, Dissertations, Professional Papers, and Capstones*; UNLV: Las Vegas, NV, USA, 2009.
49. Ren, G.; Wang, H.; Chen, C.; Wang, J. An energy conservation and environmental improvement solution-ultra-capacitor/battery hybrid power source for vehicular applications. *Sustain. Energy Technol. Assess.* **2021**, *44*, 100998. [CrossRef]
50. Coleman, M.; Lee, C.K.; Zhu, C.; Hurley, W.G. State-of-charge determination from EMF voltage estimation: Using impedance, terminal voltage, and current for lead-acid and lithium-ion batteries. *IEEE Trans. Ind. Electron.* **2007**, *54*, 2550–2557. [CrossRef]

Article

Non-Invasive Detection of Lithium-Metal Battery Degradation

Pietro Iurilli [1,*], Luigi Luppi [1,2] and Claudio Brivio [2]

1   Sustainable Energy Center, CSEM, 2002 Neuchâtel, Switzerland
2   Department of Energy, Politecnico di Milano, 20156 Milano, Italy
*   Correspondence: pietro.iurilli@csem.ch

Abstract: The application of Lithium Metal Batteries (LMBs) as secondary cells is still limited due to dendrite degradation mechanisms arising with cycling and responsible for safety risk and early cell failure. Studies to prevent and suppress dendritic growth using *state-of-the-art* materials are in continuous development. Specific detection techniques can be applied to verify the internal condition of new LMB chemistries through cycling tests. In this work, six non-invasive and BMS-triggerable detection techniques are investigated to anticipate LMB failures and to lay the basis for innovative self-healing mechanisms. The novel methodology is based on: (i) defining detection parameters to track the evolution of cell aging, (ii) defining a detection algorithm and applying it to cycling data, and (iii) validating the algorithm in its capability to detect failure. The proposed methodology is applied to Li||NMC pouch cells. The main outcomes of the work include the characterization results of the tested LMBs under different cycling conditions, the detection techniques performance evaluation, and a sensitivity analysis to identify the most performing parameter and its activation threshold.

Keywords: lithium metal batteries; dendrites; detection technique; self-healing

Citation: Iurilli, P.; Luppi, L.; Brivio, C. Non-Invasive Detection of Lithium-Metal Battery Degradation. *Energies* **2022**, *15*, 6904. https://doi.org/10.3390/en15196904

Academic Editors: Siamak Farhad and Carlos Miguel Costa

Received: 19 August 2022
Accepted: 7 September 2022
Published: 21 September 2022

Publisher's Note: MDPI stays neutral with regard to jurisdictional claims in published maps and institutional affiliations.

1. Introduction

The continuous demand for electrical storage systems with ever-growing energy density has focused research and innovation on the development of batteries that can provide superior performance over the Lithium Ions Batteries (LIBs), also called beyond-LIB technologies [1–3]. A detailed review about cathode and anode materials for developing *state-of-the-art* and future battery technologies is given by Divakaran et al. in [4], highlighting the viable materials to be used for research and development of new LIBs. Among many beyond-LIBs, rechargeable Lithium Metal Batteries (LMBs) have been extensively researched in recent years for their valuable properties [5,6]. Lithium metal is an ideal anode material for its extremely high theoretical specific capacity (3860 mAh/g), low density ($0.59 \ \mathrm{g/cm^3}$) and the lowest negative electrochemical potential (-3.05 V vs. standard hydrogen electrode). However, the applications of secondary LMBs have always been limited due to one main degradation mechanism: the dendrite growth leading to safety risk and early cell failure [7–10]. Whatever may be the testing conditions (e.g., current rate, voltage range, temperature), LMB degradation starts from the Beginning of Life (BoL) [11]. Dendrite nucleation takes place from the inhomogeneities of the Solid Electrolyte Interphase (SEI) which represents a high conductive pathway for lithium ions. The concentrated lithium ions are reduced causing fractures in the SEI layer due to volumetric expansion. These fractures represent suitable sites for deposition of lithium metal forming the dendrite [12,13]. The high conductivity of the initial deposition causes the dendritic structure to continuously grow assuming various morphologies which depend on cycling conditions [11,14]. In [15] Frenck et al. classified these morphologies in three main groups: (i) whiskers, which are long and thin needle-shaped lithium metal frames growing from the anode and covered with the SEI layer; (ii) mossy lithium, which corresponds to an accumulation of solid lithium cluster with holes and internal cavities filled with electrolyte and characterized by a large interface surface area that increases reactions with the electrolyte;

and (iii) fractal dendrites, which are thin, highly ramified structures growing directly from a layer of mossy lithium. The three listed morphologies occur in different testing conditions or after different timing that is always strictly correlated to the specific components and design of the cell under test [15]. Knowing the effects of these morphologies contribute to understanding LMB degradation. During the cell's early cycles, dendritic deposition is mossy and then progresses in a fractal configuration when the lithium ion concentration reaches zero value at the interface between the anode and the electrolyte [11,15]. The transition between mossy and fractal dendrites is described by Sand's principle and occurs at the so-called Sand's time [16]. This transition time depends on the applied current density and on the total amount of exchanged capacity. For instance, large current density leads to a fast-growing fractal dendrite structure that may penetrate the separator and may cause hazardous internal short circuits [17,18]. In addition to the short circuit risk, during the discharging phase, lithium metal is mainly stripped from dendrites causing the detachment of dendrite sections and resulting in electrically isolated inactive dead lithium [19,20]. Chen et al. deepened the dead lithium impact on the mass transport in LMBs, proving that the growth of a dead lithium layer, characterized by a small lithium ions diffusive coefficient, is associated with an increase of the electrode overpotential which results in an arcing effect of the voltage shape profile [21]. At later cycles, as the dead lithium layer thickens and the overpotential becomes more significant, the cell begins to be cycled between a much shallower voltage range, preventing a complete lithiation or delithiation of the cathode, thus resulting in cell capacity fade [22].

Studies to prevent and suppress the dendrite growth mechanisms are under continuous development and are classified as a self-healing method [23]. The most recent promising solutions involve the use of optimized electrolytes with protective additives, the modulation of lithium metal anodes and the introduction of piezoelectric separators [24–27]. Characterizing new LMB chemistries through the application of a detection technique becomes fundamental to study their behavior and to determine when dendritic degradation is going to occur in order to act accordingly. An extensive review about existing and emerging detection techniques for LMB degradation is given by Paul et al. in [28]. These detection techniques are classified into four groups, depending on their nature and application method: electrochemical in situ methods, mechanical methods, spectroscopic operando methods and chemical ex situ methods. Electrochemical in situ methods are the most promising for BMS application [29,30]. However, existing works on LMBs degradation detection are all at the development stage and have only investigated the possibility to track phenomena but not to concretely use that information to trigger actions [28].

This work analyzes and combines non-invasive detection techniques to present a novel methodology to detect LMBs degradation, laying the basis for the concrete application of LMBs self-healing mechanisms triggered by the BMS. The work is structured as follows. Section 2 shows the six selected electrochemical characterization techniques and the innovative methodology developed to apply the detection parameters as degradation identification tools. Section 3 presents the details of the experimental procedure, the most relevant results of the tests and the application of the detection techniques for the determination of approaching degradations. Section 4 discusses and identifies the most performing detection parameter based on an extensive sensitivity analysis.

2. Materials and Methods

This section includes an overview of non-invasive characterization techniques for the detection of degradation mechanisms in LMBs and the developed methodology to benchmark and validate the detection parameters extracted by the characterization techniques. The methodology has been applied to lithium metal pouch cells produced in the framework of the Horizon 2020 project HIDDEN [31]. The cells consist of Lithium Nickel Manganese Cobalt Oxide ($LiNi_{0.5}Mn_{0.3}Co_{0.2}O_2$) as the active cathode, coated on top of a 15 μm thick aluminum current collector, and a copper foil double coated with lithium metal mixed with a small amount of aluminum as an anode material (99.7% Li and 0.3% Al) and a monolayer polypropylene separator of 20–40 μm thickness. Commercially available *state-of-the-art* electrolyte for lithium metal anodes is added. A total of 5 cathode foils and 6 anode foils alternated with 12 separator layers are stacked into the pouch cell. The stack is prepared under a control environment in a glovebox and filled with the electrolyte before sealing in a four-layer pouch foil of PET, oriented nylon, aluminum and polypropylene housing. The cells have been produced in three batches at different times, with an average weight of 11 g, a theoretical capacity of 670 mAh and an estimated gravimetric energy density of 225 Wh/Kg.

2.1. Detection Techniques

As mentioned in the introduction, electrochemical non-invasive techniques that do not require cell modifications or major add-ons in a system are selected here for possible application in real BMS. Aging tracking methods, already verified for LIBs, are analyzed to select the most relevant techniques for the application with LMBs.

Incremental capacity (IC) is used as an effective method to analyze the aging mechanisms based on the discrete derivative of the cell's voltage with respect to the exchanged capacity dQ/dV [32–35]. Voltage plateaus during charging and discharging processes, corresponding to specific electrochemical phase transitions, are converted into clearly identifiable peaks on the IC curve as shown in Figure 1a. Several works proved that Loss of Active Anode Material (LAM), responsible for capacity fade and limited cyclability, leads to peak intensity reduction and peak position variation in the dQ/dV curve [36–40]. Since the formation of dead lithium causes detachment and isolation of active lithium metal, its development is directly detectable on the IC profile. Similarly to IC, *Differential Voltage* (DV) analysis is an investigation technique given by the inverse derivative dV/dQ [41–43]. Therefore, the DV curve provides information similar to the IC curve but with a different representation (peaks of IC are valleys of DV). Depending on the specific application, it may be more appropriate to use the IC or the DV profile.

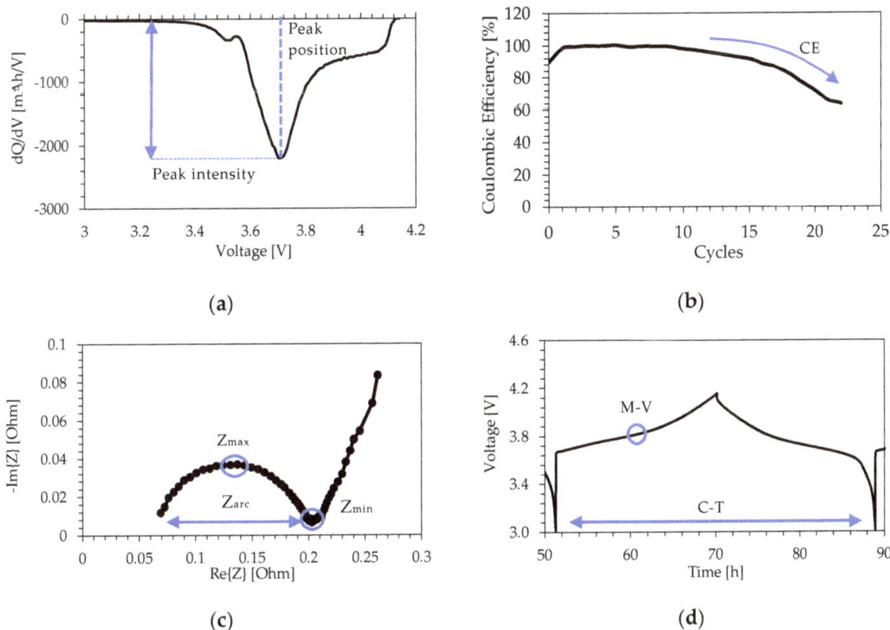

Figure 1. Characterization techniques: (**a**) Incremental Capacity (IC) during discharging phase; (**b**) Coulombic Efficiency (CE); (**c**) EIS spectrum; (**d**) voltage profile with mid-voltage (M-V) and cycle-time (C-T).

Coulombic Efficiency (CE) is a widely used technique exploited to monitor the degradation rate inside a cell [28,33,44–46]. The CE of a cell is defined as the ratio between the delivered capacity during the discharging and during the charging processes of a given cycle; an example of its trend is shown in Figure 1b. In LMBs, the formation of inactive dead lithium leads to a low CE. In [47], C. Fang et al. experimentally demonstrated that the amount of unreacted metallic lithium shows a linear relationship with loss of CE, whereas the lithium ions compound concentration is nearly constant during the CE reduction, implying that the CE decrease in LMB is led by the thickening of the isolated and unreacted lithium layer. Coulombic Efficiency Determination (CED) requires a dedicated test, consisting of continuous cycling at full Depth of Discharge (DoD). In addition, a high sampling rate is required to obtain accurate and reliable CE values, making this technique complicated to be applied and integrated into real BMS systems.

Electrochemical Impedance Spectroscopy (EIS) is a powerful characterization technique consisting of the application of a sinusoidal current signal and measuring the voltage response of the cell at different frequencies to compute the impedance [35,48–53]. Therefore, in BMS applications, a dedicated hardware is required for its application. The resulting impedance curve is normally represented in the Nyquist plot, as shown in Figure 1c and is characterized by three frequency regions [48]: high-frequency region (>1 kHz) including the inductive and ohmic behavior, mid-frequency region (0.1 Hz–1000 Hz) associated to the resistive-capacitive behavior of the charge-transfer processes and low-frequency region (<0.1 Hz) related to the capacitive behavior of the diffusive processes. As observed in [35], EIS spectra throughout cycling are affected by Loss of Lithium Inventory (LLI), that is the consumption of lithium ions by parasitic reaction (SEI growth) and decomposition reactions of the SEI or the electrolyte and LAM processes. EIS has been applied to LMBs [50], showing a large rise of the mid-frequency arch in both real and imaginary impedances and an increase of the ohmic resistance. Impedance evolution during cell aging can be tracked by fitting the curve with Equivalent Circuit Models (ECMs) and by tracking the impedance value of specific parameters (e.g., the maximum of the arch, the width of the arch and the minimum before the tail) [54,55].

Mid-Voltage (M-V) and Cycle-Time (C-T): As a consequence of the dead lithium layer thickening, an increasing electrode overpotential is introduced, leading to an arcing in the voltage profile during the charging process [21]. Furthermore, the increasing overpotential causes the voltage to reach the set maximum and the minimum limit values faster in time during cycling. The evolution of the voltage shape profile (M-V) is recorded by measuring the voltage value at the halfway point in time during charging, as represented in Figure 1d. The time duration of a cycle is then adopted to evaluate the aging state of the cell and the ongoing impact of dead lithium with the C-T parameter.

Table 1 summarizes and ranks the listed characterization techniques depending on different criteria. EIS can be performed in the shortest time. All detection techniques can be implemented into embedded systems (e.g., BMS), but both EIS and CE are considered more complex measurement techniques. IC/DV and EIS parameters allow for the detection of degradation and discriminate the degradation modes (e.g., loss of active material, loss of lithium ions). Mid-Voltage and Cycle Time parameters only allow us to detect the accumulation of dead lithium without more details. Finally, some of the listed techniques allow us to calculate Li plating and stripping rates. These values are useful to quantify the cell degradation in more detail.

2.2. Methodology

The procedure developed to benchmark the detection techniques is schematically represented in Figure 2, including four steps: cell testing, detection parameters identification, thresholds setting and validation. These steps will be described in the next paragraphs.

Figure 2. Schematic representation of the developed methodology.

Table 1. Evaluation of measurement and detection characteristics of the techniques analyzed for LMB degradation detection.

Technique	Length	Complexity	Applicability	Degradation Modes	Dendrite Quantification
IC	■■□	■□□	■■□	■■■	■□□
DV	■■□	■□□	■■□	■■■	■■■
CE	■■■	■■□	□□□	■■□	■■■
EIS	■□□	■■□	■■□	■■■	■■□
M–V	■■□	■□□	■■□	■■□	□□□
C–T	■■□	■□□	■■□	■■□	□□□

2.2.1. Cell Testing

A scheme of the cell testing protocol is represented in Figure 3. The procedure is divided in three main steps:

- Formation and characterization: The formation process for the initial SEI layer establishment is performed via two full cycles at low C-rate. The characterization is based on the Battery Capacity Determination (BCD)—a full charge and discharge cycle at specific current rate—and EIS measurements.
- Cycling: consecutive charge–discharge cycles at specific C-rate and DoD.
- Diagnosis: The cell properties are checked regularly after a fixed number of cycles. BCD is performed to obtain the cell's updated capacity (i.e., SoH) and to compute the IC, M-V and C-T characterization techniques. EIS in the frequency range of 10 kHz–10 mHz with C/50 AC current excitation are performed at different OCVs (i.e., SoCs).

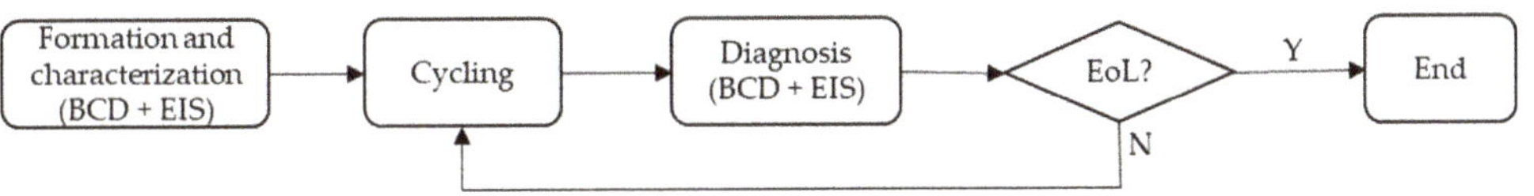

Figure 3. Schematic representation of the cell testing protocol.

After the initial characterization step, cycling and diagnosis are lopped until the cell reaches the End of Life (EoL), represented by a SoH lower than 70%. The only exception is represented by the Coulombic Efficiency protocol, which consists of continuous cycling without interruptions until the EoL of the cell.

2.2.2. Detection Parameters Identification

The experimental results collected during the cell testing are analyzed to define the detection parameters as reported in the last column of Table 2:

- *IC*: The variation of the peak intensity during the discharge phase constitutes a parameter to follow the degradation of the cell. Generally, low current rates during the diagnosis cycle are recommended to have the cell close to equilibrium condition and thus obtain better parameter results. Low rates are preferable (~C/20), but it is proven that even at higher rates (~1C), relevant aging signs are still identifiable [39]. Since the information of the DV and IC profile are similar, only IC analysis is adopted in this study.

- *CE*: The CE technique itself represents the detection parameter.
- *EIS*: Impedance spectra are tracked by recording specific impedance in the three main frequency zones: the local maximum, the width of the mid-frequency semiarch, and the local minimum before the diffusive tail. The resulting Z_{max} Z_{arch}, and Z_{min} represent suitable parameters to track the impedance evolution as the cell faces aging processes. In this study, the imaginary part of Z_{max} and Z_{min}, is used.
- *M-V and C-T*: themselves represent the parameters for detection. Both the parameters are retrieved from full diagnosis cycles at regular intervals during testing.

Table 2. Description of characterization techniques and overview of selected detection parameters.

Technique	Measurement Protocol	Elaboration (If Present)	Detection Parameter
IC	Full charge or discharge	$IC = \frac{dQ}{dV}$	Peak intensity
CE	Full cycle	$CE = \frac{Q_{disch}}{Q_{charge}}$	CE value
EIS	EIS	$Z_{arch} = Z_{min} - Z_{ohm}$	Z_{arch}, Z_{max_im}, Z_{min_im}
M–V	Full charge	$Mid - voltage = V_{charge}(t = \frac{t_{charge}}{2})$	M-V
C-T	Full cycle	$Cycle\ time = t_{end} - t_0$	C-T

2.2.3. Thresholds Setting

Through the application of the above presented techniques, the detection parameters are extracted and tracked. Figure 4a exemplifies a parameter variation with respect to its initial value. The parameters usually exhibit small variations between consecutive diagnostic assessments, whereas large and rapid changes occur at later cycles when degradation processes become more relevant. A threshold is set to identify when the transition between the low and high degradation rate occurs:

$$\Delta p = p_i - p_{i-1} > [\text{Threshold}] \tag{1}$$

where Δp is the difference between the actual parameter value p_i and its previous point p_{i-1}. If a threshold is exceeded during testing, as represented at the check n.3 of Figure 4a, a warning signal is triggered to report an upcoming degradation.

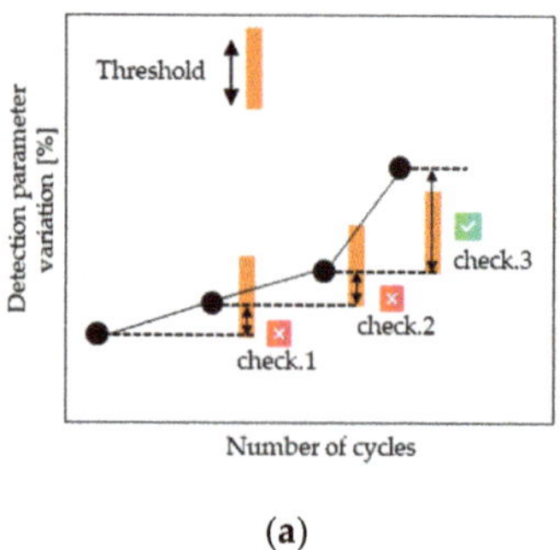

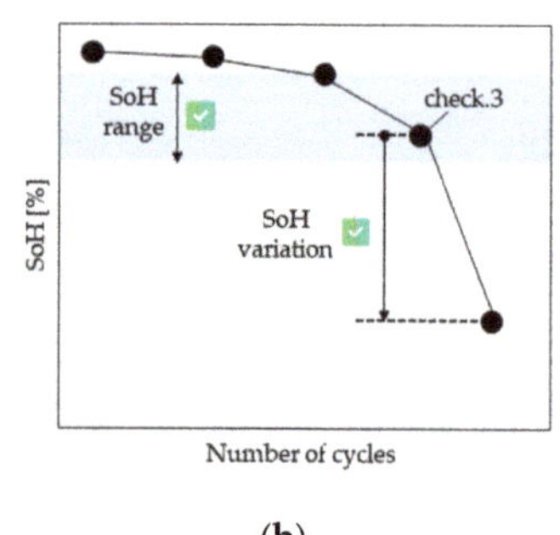

(a) (b)

Figure 4. Detection algorithm: (**a**) application of thresholds between consecutive measurements of detection parameters; (**b**) validation procedure.

2.2.4. Validation

The validation procedure is then applied to verify whether the warning signal resulting from exceeding a specific parameter threshold truly represents a condition preceding a degradation of the LMB. The process is based on two State of Health (SoH) checks at the triggering event:

- SoH range: At the time when the trigger is activated, the SoH of the cell must be between two limits. This is performed to avoid too early or too late self-healing activation.
- SoH variation: The variation between the SoH of the cell at the time when the trigger is activated and the SoH at the following diagnosis step is used to check if the parameter is detecting and preventing a fast degradation.

Figure 4b shows a graphical representation of the validation procedure for the generic parameter triggered at the check n.3. The fulfillment of both criteria implies that the detection parameter with the selected threshold is suitable for the LMB failure detection.

3. Results and Discussion

Characterization and cycling tests were performed at the Sustainable Energy Center laboratories of CSEM in Neuchâtel, Switzerland with a BioLogic BCS 815 [56] battery tester ($\pm$0.01% FSD accuracy on the voltage, $\pm$0.015% FSD accuracy on current, for each available range and EIS capability from 10 kHz to 10 mHz). The cells were placed inside a thermostatic chamber Angelantoni ATT-DM340 [57] to prevent large temperature variations. Table 3 collects the main characteristics of the tests performed on the three batches, under the testing protocol shown in Figure 3: (i) number of cycles performed, the C-rate and voltage range during formation; (ii) capacity determination and EIS during first characterization, (iii) C-rate, DoD and number of cycles applied during cycling phase; and (iv) capacity determination and EIS during the diagnosis phase. In short, the different batches focused on different aspects:

- 1st batch (C-rate focused): The cells are cycled at different C-rates and at full cell capacity (100% DoD), performing a complete diagnosis (BCD+EIS) every 10 cycles.
- 2nd batch (DoD focused): The cells are cycled at reduced DoD, keeping the same C-rate throughout the whole test. EIS during charging are performed every equivalent cycle.
- 3rd batch (hybrid): The tests alternate cycling phases at reduced DoD and full diagnosis every five equivalent cycles.

Table 3. Testing protocol specifications for the three different batches of cells.

| | Formation | 1st Characterization | Cycling | | | | Diagnosis |
			Cell ID	C-Rate	DoD	n. of Cycles	
1st batch	3 cycles, C/10, 3–4.3 V	BCD at C/10, EIS at different SoCs	5	C/10	100%	10	BCD at C/10, EIS at different SoCs
			6	C/5	100%	10	
			7	C/2	100%	10	
			8	1C	100%	10	
			10	C/10	100%	-	CED
			11	C/5	100%	-	CED
			12	C/2	100%	-	CED
2nd batch	2 cycles, C/20, 3–4 V	EIS at 3.8V	15	C/20	20%	5	EIS at 3.8 V
			21	C/20	50%	2	
			22	C/20	80%	1	
			24	C/20	100%	1	
			19	C/10	100%	1	
			17	C/20	100%	-	CED
3rd batch	2 cycles, C/20, 3–4 V	BCD at C/20, EIS at different SoCs	36	C/20	20%	25	BCD at C/10, EIS at different SoCs
			29	C/20	50%	10	
			38	C/20	80%	5	
			37	C/20	100%	5	

The cells that performed Coulombic Efficiency Determination are labelled as "CED" in the column diagnosis (first batch and second batch).

3.1. Testing Results

When analyzing the most relevant results, Figure 5a,b show the SoH evolution retrieved from the BCD diagnosis as a function of the equivalent cycles, respectively, for the 1st and the 3rd batches. EIS spectra evolution (Figure 5c) shows a great rise both in real and imaginary parts as the cell ages. Similarly, the IC profiles (Figure 5d) show a main peak around 3.7 V that consistently decreases its intensity during aging and shifts rightward between the third and fourth diagnosis steps.

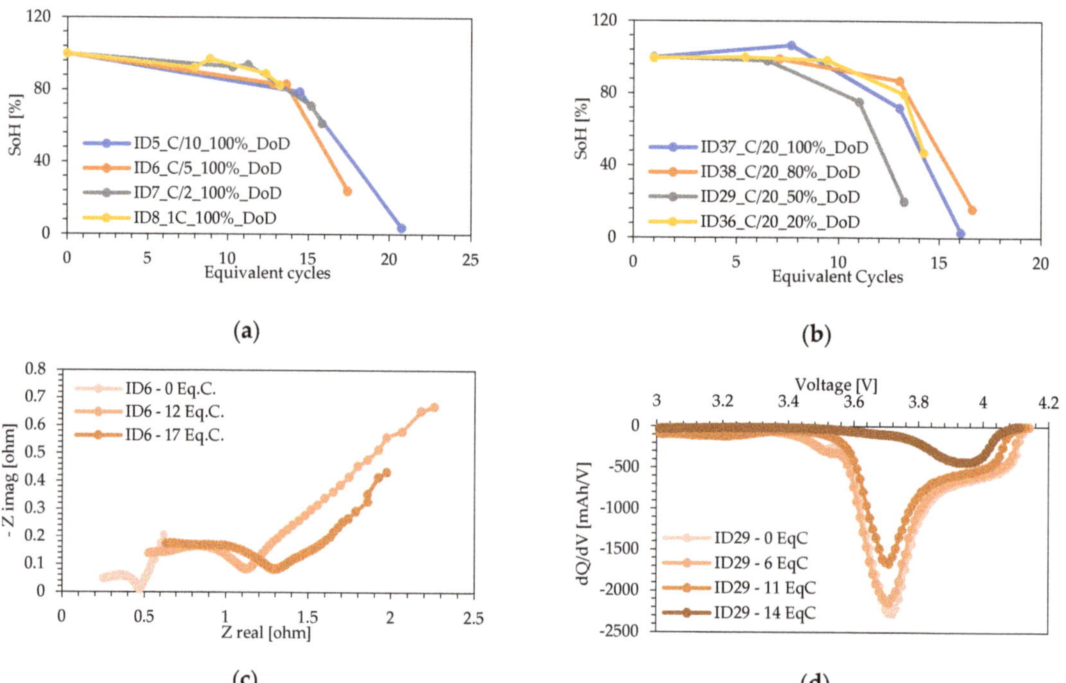

Figure 5. Test results: (**a**) SoH evolution profiles of the 1st batch cells; (**b**) SoH evolution profiles of the 3rd batch cells; (**c**) EIS spectra evolution of cell ID6 at 3.9 V; (**d**) IC profiles of cell ID29.

Figure 6 shows the trends of the detection parameters listed in Table 2 for the 3rd batch cells. For instance, Figure 6a shows the IC peak intensity variation batch, Figure 6b represents the M-V parameter variation and Figure 6c shows the evolution of Z_{arch}. All the parameters are evaluated with respect to their initial value at BoL. The parameters follow a common trend characterized by a sudden increase at the 12th equivalent cycle: this growth is representative of a significant change in cell behavior, and it corresponds to the strong SoH fade represented in Figure 5b.

Lastly, CED is analyzed in Figure 7a,b. The higher the current rate is, the earlier the steep discharge capacity fade occurs, and consequently, the earlier CE drops to values lower than 100%. The irregular trend of cell ID12 suggests that a current rate of C/2 is too high for the cells under investigation.

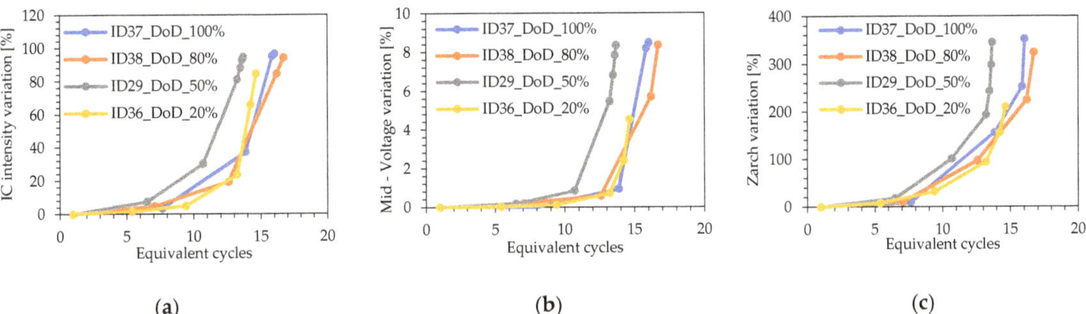

Figure 6. Examples of detection parameters trends over equivalent cycles for the cells of the 3rd batch: (**a**) IC peak intensity; (**b**) M-V, (**c**) Z_{arch}.

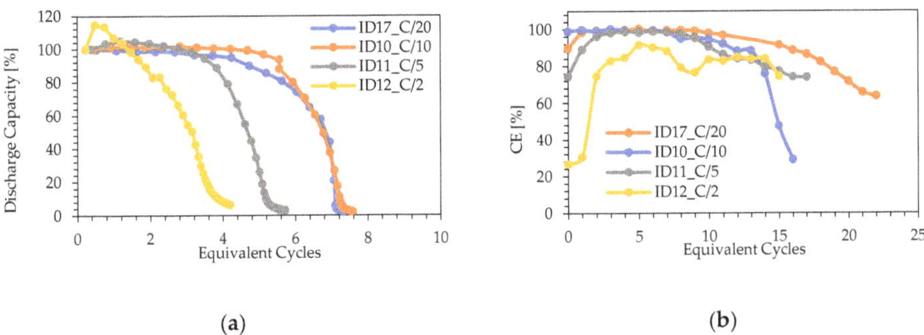

Figure 7. CED test results: (**a**) evolution of discharge capacity over equivalent cycles with respect to the initial value; and (**b**) CE trends of the tested cells over the equivalent cycles.

Overall, the collected parameters show clear and distinct signals of cell degradation, i.e., internal physical changes. Given the arcing effect observed in cells voltage profiles during cycling, the degradation is mostly attributed to an accumulation of dead lithium [21]. This phenomenon causes a sudden and steep variation of the detection parameter occurring just after an approximately steady condition.

3.2. Triggering Thresholds and Detection Parameters Validation

As per the procedure detailed in Figure 2, for each of the selected parameters, a suitable threshold has been empirically defined as variation between two consecutive diagnosis steps. Table 4 shows the selected values. In the case of M-V and C-T, a second limit has been introduced to improve reliability of these indicators. The detection techniques with the selected thresholds have been applied to the three batches, with the most representative results reported in Figure 8.

Table 4. Triggering thresholds chosen for the detection parameters investigated.

Parameter	Threshold Values
Mid-Voltage	$(M - V_i) - (M - V_{i-1}) > 1.25\%$ $(M - V_i) - (M - V_0) > 2.5\%$
Cycle Time	$(C - T_i) - (C - T_{i-1}) < 4\%$ $(C - T_i) < 90\%$
Incremental Capacity	$Peak\_Intensity_i - Peak\_intensity_{i-1} > 5\%$
Z_{max_im}	$Z_{max\_i} - Z_{max\_i-1} > 15\%$
Z_{max_im}	$Z_{min\_i} - Z_{min\_i-1} > 15\%$
Z_{arch}	$Z_{arch\_i} - Z_{arch\_i-1} > 15\%$
Coulombic Efficiency	$CE_i - CE_{i-1} > 1.5\%$

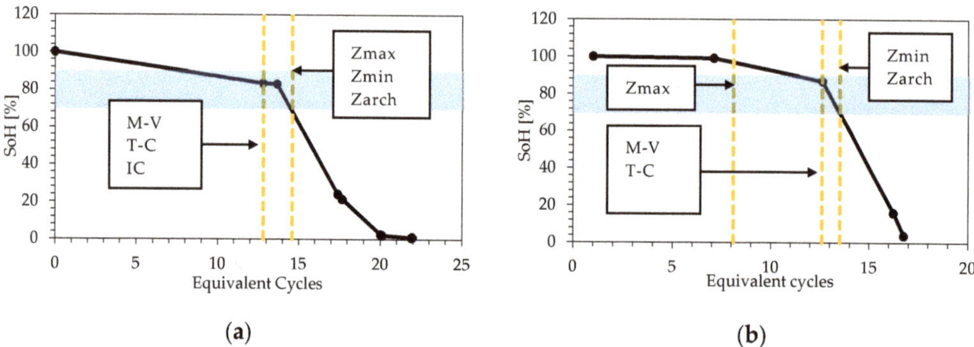

Figure 8. Detection results with triggering highlighted in yellow for: (**a**) cell ID6 of the 1st batch cycling at C/5; (**b**) cell ID38 of the 3rd batch cycling at 80% DoD and C/20.

Regarding the 1st batch experiment, Figure 8a shows that the degradation detection for cell ID6 is triggered in proximity of the SoH slope change for all the tested techniques. Voltage profile-based parameters (i.e., IC, M-V, C-T) showed an earlier detection than impedance-based ones, with the triggering event at the 13th EqC. Similar results were obtained for cell ID38 (third batch), as shown in Figure 8b, with some early triggering by the Z_{max} parameter around the eighth cycle. Analogously, the detection parameters have been evaluated for all the other tested cells. The degradation detection is validated whenever the two criteria presented in Section 2.2.4 are fulfilled, specifically: (i) 70% < SoH_i < 90% and (ii) $SoH_{i+1} - SoH_i \geq 5\%$. Table 5 shows the validation results for the three batches. Unfortunately, the 1st batch has been affected by the limited number of diagnostic phases in the tests. In the case of the 2nd and 3rd batches, M-V, C-T and IC peak intensity parameters fulfill both validation criteria in almost all the testing conditions suggesting a good performance of the parameters to detect degradation.

Table 5. Validation procedure results for the three batch tests. The columns correspond to the testing conditions and the rows to the detection parameters. The legend is given in the lower part of the table.

Parameter	1st Batch					2nd Batch					3rd Batch		
	C/10—100% DoD	C/5—100% DoD	C/2—100% DoD	1C—100% DoD	C/20—100% DoD	C/10—100% DoD	C/20—80% DoD	C/20—50% DoD	C/20—20% DoD	C/20—100% DoD	C/20—80% DoD	C/20—50% DoD	C/20—20% DoD
Z_{max_im}	√|√	√|√	×|×	×|×	×|×	√|√	√|√	√|√	×|√	√|×	×|×	×|×	×|×
Z_{min_im}	√|√	√|√	×|×	×|×	√|√	√|√	√|√	√|√	√|×	√|×	√|√	×|√	×|√
Z_{arch}	√|√	√|√	×|×	×|×	√|√	√|√	√|√	√|√	×|√	√|×	√|√	×|×	×|×
M-V	√|√	√|√	×|×	×|×	×|×	×|√	√|√	×|√	√|×	√|√	√|√	√|√	√|√
C-T	√|√	√|√	×|×	√|√	√|×	√|√	×|√	√|√	√|×	√|√	√|√	√|√	√|√
IC peak intensity	√|√	√|√	×|×	×|×	×|×	√|√	×|√	√|√	√|√	√|√	√|√	×|×	√|√
CE	×|√	√|√	√|√	–	√|√	–	–	–	–	–	–	–	–
Legend	×|×: both SoH checks failed						√|×: SoH range fulfilled; SoH variation failed						
	√|√: both SoH checks fulfilled						×|√: SoH range failed; SoH variation fulfilled						

3.3. Sensitivity Analysis

A sensitivity analysis was applied to identify the best values for the parameters' thresholds (Table 6). Ten different thresholds values were tested for each detection parameter. The sensitivity analysis was applied only to the 2nd and 3rd batches, given the few diagnosis points available in the 1st batch. The thresholds are evaluated separately, and the detection Success Rate was computed, i.e., the ratio between the number of cases where the SoH validation criterium was fulfilled over the total number of tested cases.

Table 6. Success rate results for SoH range and SoH variation validation procedures with the parameter thresholds selected for the sensitivity analysis application.

	Thresholds	2.5%	5%	7.5%	10%	12.5%	15%	17.5%	20%	22.5%	25%
Z_{max_im} [%]	SoH range	22%	22%	22%	33%	33%	33%	22%	44%	22%	22%
	SoH var.	44%	44%	56%	56%	56%	56%	56%	78%	67%	78%
	Thresholds	2.5%	5%	7.5%	10%	12.5%	15%	17.5%	20%	22.5%	25%
Z_{min_im} [%]	SoH range	22%	22%	22%	44%	56%	67%	67%	67%	56%	33%
	SoH var.	44%	44%	44%	67%	78%	89%	89%	89%	100%	78%
	Thresholds	2.5%	5%	7.5%	10%	12.5%	15%	17.5%	20%	22.5%	25%
Z_{arch} [%]	SoH range	22%	33%	33%	33%	44%	44%	44%	56%	33%	22%
	SoH var.	44%	44%	44%	56%	67%	67%	67%	89%	78%	67%
	Thresholds	0.1%	0.25%	0.5%	0.75%	1%	1.25%	1.5%	2%	2.25%	2.5%
M-V [%]	SoH range	33%	33%	44%	67%	67%	67%	56%	44%	44%	44%
	SoH var.	44%	44%	56%	78%	78%	78%	67%	78%	67%	67%
	Thresholds	1%	1.5%	2%	2.5%	3%	4%	5%	6%	7.5%	10%
C-T [%]	SoH range	33%	44%	56%	56%	78%	89%	100%	100%	100%	100%
	SoH var.	56%	56%	67%	67%	78%	78%	78%	78%	78%	78%
	Thresholds	0.25%	0.5%	0.75%	1%	2.5%	5%	7.5%	10%	12.5%	15%
IC [%]	SoH range	11%	11%	11%	11%	22%	56%	67%	67%	67%	56%
	SoH var.	44%	44%	44%	44%	44%	78%	89%	100%	89%	67%
	Thresholds	0.1%	0.25%	0.5%	1%	1.5%	2%	2.5%	3%	4%	5%
CE [%]	SoH range	25%	25%	25%	50%	50%	50%	50%	50%	50%	25%
	SoH var.	0%	0%	25%	25%	75%	50%	50%	50%	75%	50%

The resulting Success Rates are reported in Table 6, respectively, for the SoH range and the SoH variation validation criteria. In the first case, mid-range thresholds allow for a higher success rate applying all the detection techniques. Too small or too high threshold values are ineffective, respectively, with too early and too late detection. In the second case, the results for the SoH variation are very similar, even though larger threshold values give a higher success rate for most of the techniques.

Thresholds' Success Rates for the two validation criteria were then combined by retaining the lowest Success Rate value among the two applied criteria. The best threshold value is defined as the one with the highest success rate for a specific detection parameter. The results are listed in Table 7. Overall, the Cycle Time obtained the highest success rate of 78%, the Mid-Voltage and IC peak intensity share the same success rate of 67%, whereas the best performing impedance-based parameter is Z_{min} with a 67% success rate, followed by Z_{arch} and finally Z_{max}. Lastly, the CE parameter led to effective detection in only half of the tested cases.

Table 7. Sensitivity analysis results.

Parameter	Threshold Value [%]	Success Rate [%]
Z_{max_im}	20%	44%
Z_{min_im}	15%	67%
Z_{arch_im}	20%	56%
M-V	0.75%	67%
C-T	3%	78%
IC	7.5%	67%
CE	1.5%	50%

4. Conclusions

This work introduced a structured methodology to test and validate parameters that can be used to detect degradation in LMB and which should be non-invasive and BMS-triggerable. In this way, a novel algorithm structure is presented to enable LMB degradation tracking and to trigger self-healing mechanism activation by BMS. After a review of the selected electrochemical techniques (Section 1), Section 2 presented the developed methodology which includes (i) the cycling and diagnosis of standard LMBs via six testing protocols, (ii) the analysis on the selected detection parameters trends with respect to capacity fade, (iii) the thresholds definition to trigger a self-healing action and (iv) the performance validation in early detecting LMBs degradation. Section 3 presented the case study where three batches of LMBs have been tested with different protocol focus: different C-rates and 100% DoD, fixed C-rate and different DoD ranges. A sensitivity analysis was performed to compute the Success Rate of each detection technique. On average, the detection techniques successfully detected degradation in 60% of the cases and can be used to anticipate LMB failures and to lay the basis for self-healing mechanism activation.

Limitations of this work can be analyzed on different levels. *Validity of the results*: The obtained Success Rate was not directly linked to the ability to detect dendrite growth formation. During the experimental phase, it was in fact not possible to discriminate dendrite growth from dead lithium, the fast accumulation of which was confirmed by post-mortem analysis. All the tested cells reached their EoL due to cycling failure and not for short-circuit failure. *Replicability of the results*: The obtained Success Rates are very dependent on the experimental campaign performed for two reasons: statistics, due to the limited number of tested samples, and specificities of the tested cells (Li | | NMC). The developed methodology can be replicated to other LMB with different chemical formulation, but different performance results should be expected. These results could also be influenced by more severe or real use profiles testing conditions. *Applicability of the results*: The tested detection techniques can be implemented in embedded systems to sense degradation and trigger actions such as, for instance, self-healing methods, which is one of the possible routes to make LMB exploitable in real applications. Some techniques are easier than others, but in general all the non-invasive techniques proposed can be implemented in BMSs.

Author Contributions: Conceptualization, P.I. and C.B.; methodology, L.L. and P.I.; validation, L.L.; investigation, L.L. and P.I.; writing—original draft preparation, L.L.; writing—review and editing, P.I. and C.B.; visualization, P.I.; supervision, C.B.; funding acquisition, C.B. All authors have read and agreed to the published version of the manuscript.

Funding: This research was funded by the HIDDEN project that receives funding from the European Union's Horizon 2020 research and innovation programme under Grant Agreement No. 957202.

Data Availability Statement: The data presented in this study are openly available in Zenodo at 10.5281/zenodo.7007544.

Acknowledgments: The authors would like to thank Marco Merlo, Energy Department, Politecnico di Milano, Milano (Italy) for his support in developing the testing methodology.

Conflicts of Interest: The authors declare no conflict of interest.

References

1. Tian, Y.; Zeng, G.; Rutt, A.; Shi, T.; Kim, H.; Wang, J.; Koettgen, J.; Sun, Y.; Ouyang, B.; Chen, T.; et al. Promises and Challenges of Next-Generation "Beyond Li-Ion" Batteries for Electric Vehicles and Grid Decarbonization. *Chem. Rev.* **2021**, *121*, 1623–1669. [CrossRef]
2. Shen, X.; Liu, H.; Cheng, X.-B.; Yan, C.; Huang, J.-Q. Beyond Lithium Ion Batteries: Higher Energy Density Battery Systems Based on Lithium Metal Anodes. *Energy Storage Mater.* **2018**, *12*, 161–175. [CrossRef]
3. Thackeray, M.M.; Wolverton, C.; Isaacs, E.D. Electrical Energy Storage for Transportation—Approaching the Limits of, and Going beyond, Lithium-Ion Batteries. *Energy Environ. Sci.* **2012**, *5*, 7854. [CrossRef]
4. Divakaran, A.M.; Minakshi, M.; Bahri, P.A.; Paul, S.; Kumari, P.; Divakaran, A.M.; Manjunatha, K.N. Rational Design on Materials for Developing next Generation Lithium-Ion Secondary Battery. *Prog. Solid State Chem.* **2021**, *62*, 100298. [CrossRef]
5. Wang, R.; Cui, W.; Chu, F.; Wu, F. Lithium Metal Anodes: Present and Future. *J. Energy Chem.* **2020**, *48*, 145–159. [CrossRef]
6. Wang, Z.; Cao, Y.; Zhou, J.; Liu, J.; Shen, X.; Ji, H.; Yan, C.; Qian, T. Processing Robust Lithium Metal Anode for High-Security Batteries: A Minireview. *Energy Storage Mater.* **2022**, *47*, 122–133. [CrossRef]
7. Aurbach, D. A Short Review of Failure Mechanisms of Lithium Metal and Lithiated Graphite Anodes in Liquid Electrolyte Solutions. *Solid State Ion.* **2002**, *148*, 405–416. [CrossRef]
8. Lu, D.; Shao, Y.; Lozano, T.; Bennett, W.D.; Graff, G.L.; Polzin, B.; Zhang, J.; Engelhard, M.H.; Saenz, N.T.; Henderson, W.A.; et al. Failure Mechanism for Fast-Charged Lithium Metal Batteries with Liquid Electrolytes. *Adv. Energy Mater.* **2015**, *5*, 1400993. [CrossRef]
9. Zhang, Y.; Zhong, Y.; Shi, Q.; Liang, S.; Wang, H. Cycling and Failing of Lithium Metal Anodes in Carbonate Electrolyte. *J. Phys. Chem. C* **2018**, *122*, 21462–21467. [CrossRef]
10. Raj, V.; Aetukuri, N.P.B.; Nanda, J. Solid State Lithium Metal Batteries—Issues and Challenges at the Lithium-Solid Electrolyte Interface. *Curr. Opin. Solid State Mater. Sci.* **2022**, *26*, 100999. [CrossRef]
11. Wood, K.N.; Noked, M.; Dasgupta, N.P. Lithium Metal Anodes: Toward an Improved Understanding of Coupled Morphological, Electrochemical, and Mechanical Behavior. *ACS Energy Lett.* **2017**, *2*, 664–672. [CrossRef]
12. Wood, K.N.; Kazyak, E.; Chadwick, A.F.; Chen, K.-H.; Zhang, J.-G.; Thornton, K.; Dasgupta, N.P. Dendrites and Pits: Untangling the Complex Behavior of Lithium Metal Anodes through Operando Video Microscopy. *ACS Cent. Sci.* **2016**, *2*, 790–801. [CrossRef]
13. Stark, J.K.; Ding, Y.; Kohl, P.A. Nucleation of Electrodeposited Lithium Metal: Dendritic Growth and the Effect of Co-Deposited Sodium. *J. Electrochem. Soc.* **2013**, *160*, D337–D342. [CrossRef]
14. Horstmann, B.; Shi, J.; Amine, R.; Werres, M.; He, X.; Jia, H.; Hausen, F.; Cekic-Laskovic, I.; Wiemers-Meyer, S.; Lopez, J.; et al. Strategies towards Enabling Lithium Metal in Batteries: Interphases and Electrodes. *Energy Environ. Sci.* **2021**, *14*, 5289–5314. [CrossRef]
15. Frenck, L.; Sethi, G.K.; Maslyn, J.A.; Balsara, N.P. Factors That Control the Formation of Dendrites and Other Morphologies on Lithium Metal Anodes. *Front. Energy Res.* **2019**, *7*, 115. [CrossRef]
16. Bai, P.; Li, J.; Brushett, F.R.; Bazant, M.Z. Transition of Lithium Growth Mechanisms in Liquid Electrolytes. *Energy Environ. Sci.* **2016**, *9*, 3221–3229. [CrossRef]
17. Kong, L.; Xing, Y.; Pecht, M.G. In-Situ Observations of Lithium Dendrite Growth. *IEEE Access* **2018**, *6*, 8387–8393. [CrossRef]
18. Rosso, M.; Brissot, C.; Teyssot, A.; Dollé, M.; Sannier, L.; Tarascon, J.-M.; Bouchet, R.; Lascaud, S. Dendrite Short-Circuit and Fuse Effect on Li/Polymer/Li Cells. *Electrochim. Acta* **2006**, *51*, 5334–5340. [CrossRef]
19. Xu, S.; Chen, K.-H.; Dasgupta, N.P.; Siegel, J.B.; Stefanopoulou, A.G. Evolution of Dead Lithium Growth in Lithium Metal Batteries: Experimentally Validated Model of the Apparent Capacity Loss. *J. Electrochem. Soc.* **2019**, *166*, A3456–A3463. [CrossRef]
20. Aryanfar, A.; Brooks, D.J.; Colussi, A.J.; Hoffmann, M.R. Quantifying the Dependence of Dead Lithium Losses on the Cycling Period in Lithium Metal Batteries. *Phys. Chem. Chem. Phys.* **2014**, *16*, 24965–24970. [CrossRef]
21. Chen, K.-H.; Wood, K.N.; Kazyak, E.; LePage, W.S.; Davis, A.L.; Sanchez, A.J.; Dasgupta, N.P. Dead Lithium: Mass Transport Effects on Voltage, Capacity, and Failure of Lithium Metal Anodes. *J. Mater. Chem. A* **2017**, *5*, 11671–11681. [CrossRef]
22. Niu, C.; Lee, H.; Chen, S.; Li, Q.; Du, J.; Xu, W.; Zhang, J.-G.; Whittingham, M.S.; Xiao, J.; Liu, J. High-Energy Lithium Metal Pouch Cells with Limited Anode Swelling and Long Stable Cycles. *Nat. Energy* **2019**, *4*, 551–559. [CrossRef]
23. Narayan, R.; Laberty-Robert, C.; Pelta, J.; Tarascon, J.-M.; Dominko, R. Self-Healing: An Emerging Technology for Next-Generation Smart Batteries. *Adv. Energy Mater.* **2022**, *12*, 2102652. [CrossRef]
24. Hu, Z.; Li, G.; Wang, A.; Luo, J.; Liu, X. Recent Progress of Electrolyte Design for Lithium Metal Batteries. *Batter. Supercaps* **2020**, *3*, 331–335. [CrossRef]
25. Ma, L.; Cui, J.; Yao, S.; Liu, X.; Luo, Y.; Shen, X.; Kim, J.-K. Dendrite-Free Lithium Metal and Sodium Metal Batteries. *Energy Storage Mater.* **2020**, *27*, 522–554. [CrossRef]
26. Gao, T.; Rainey, C.; Lu, W. Piezoelectric Mechanism and a Compliant Film to Effectively Suppress Dendrite Growth. *ACS Appl. Mater. Interfaces* **2020**, *12*, 51448–51458. [CrossRef]
27. Liu, G.; Wang, D.; Zhang, J.; Kim, A.; Lu, W. Preventing Dendrite Growth by a Soft Piezoelectric Material. *ACS Mater. Lett.* **2019**, *1*, 498–505. [CrossRef]
28. Paul, P.P.; McShane, E.J.; Colclasure, A.M.; Balsara, N.; Brown, D.E.; Cao, C.; Chen, B.; Chinnam, P.R.; Cui, Y.; Dufek, E.J.; et al. A Review of Existing and Emerging Methods for Lithium Detection and Characterization in Li-Ion and Li-Metal Batteries. *Adv. Energy Mater.* **2021**, *11*, 2100372. [CrossRef]

29. Konz, Z.M.; McShane, E.J.; McCloskey, B.D. Detecting the Onset of Lithium Plating and Monitoring Fast Charging Performance with Voltage Relaxation. *ACS Energy Lett.* **2020**, *5*, 1750–1757. [CrossRef]

30. Attia, P.M.; Grover, A.; Jin, N.; Severson, K.A.; Markov, T.M.; Liao, Y.-H.; Chen, M.H.; Cheong, B.; Perkins, N.; Yang, Z.; et al. Closed-Loop Optimization of Fast-Charging Protocols for Batteries with Machine Learning. *Nature* **2020**, *578*, 397–402. [CrossRef]

31. The HIDDEN Project. Available online: https://hidden-project.eu/ (accessed on 21 March 2022).

32. Berecibar, M.; Dubarry, M.; Omar, N.; Villarreal, I.; Van Mierlo, J. Degradation Mechanism Detection for NMC Batteries Based on Incremental Capacity Curves. *World Electr. Veh. J.* **2016**, *8*, 350–361. [CrossRef]

33. Yang, F.; Wang, D.; Zhao, Y.; Tsui, K.-L.; Bae, S.J. A Study of the Relationship between Coulombic Efficiency and Capacity Degradation of Commercial Lithium-Ion Batteries. *Energy* **2018**, *145*, 486–495. [CrossRef]

34. Li, X.; Yuan, C.; Li, X.; Wang, Z. State of Health Estimation for Li-Ion Battery Using Incremental Capacity Analysis and Gaussian Process Regression. *Energy* **2020**, *190*, 116467. [CrossRef]

35. Pastor-Fernández, C.; Uddin, K.; Chouchelamane, G.H.; Widanage, W.D.; Marco, J. A Comparison between Electrochemical Impedance Spectroscopy and Incremental Capacity-Differential Voltage as Li-Ion Diagnostic Techniques to Identify and Quantify the Effects of Degradation Modes within Battery Management Systems. *J. Power Sources* **2017**, *360*, 301–318. [CrossRef]

36. Krupp, A.; Ferg, E.; Schuldt, F.; Derendorf, K.; Agert, C. Incremental Capacity Analysis as a State of Health Estimation Method for Lithium-Ion Battery Modules with Series-Connected Cells. *Batteries* **2020**, *7*, 2. [CrossRef]

37. Dubarry, M.; Svoboda, V.; Hwu, R.; Liaw, B.Y. Incremental Capacity Analysis and Close-to-Equilibrium OCV Measurements to Quantify Capacity Fade in Commercial Rechargeable Lithium Batteries. *Electrochem. Solid-State Lett.* **2006**, *9*, A454. [CrossRef]

38. Anseán, D.; García, V.M.; González, M.; Blanco-Viejo, C.; Viera, J.C.; Pulido, Y.F.; Sánchez, L. Lithium-Ion Battery Degradation Indicators Via Incremental Capacity Analysis. *IEEE Trans. Ind. Appl.* **2019**, *55*, 2992–3002. [CrossRef]

39. Plattard, T.; Barnel, N.; Assaud, L.; Franger, S.; Duffault, J.-M. Combining a Fatigue Model and an Incremental Capacity Analysis on a Commercial NMC/Graphite Cell under Constant Current Cycling with and without Calendar Aging. *Batteries* **2019**, *5*, 36. [CrossRef]

40. Schmitt, J.; Schindler, M.; Oberbauer, A.; Jossen, A. Determination of Degradation Modes of Lithium-Ion Batteries Considering Aging-Induced Changes in the Half-Cell Open-Circuit Potential Curve of Silicon–Graphite. *J. Power Source* **2022**, *532*, 231296. [CrossRef]

41. Kato, H.; Kobayashi, Y.; Miyashiro, H. Differential Voltage Curve Analysis of a Lithium-Ion Battery during Discharge. *J. Power Source* **2018**, *398*, 49–54. [CrossRef]

42. Bloom, I.; Walker, L.K.; Basco, J.K.; Abraham, D.P.; Christophersen, J.P.; Ho, C.D. Differential Voltage Analyses of High-Power Lithium-Ion Cells. 4. Cells Containing NMC. *J. Power Source* **2010**, *195*, 877–882. [CrossRef]

43. Keil, P.; Jossen, A. Calendar Aging of NCA Lithium-Ion Batteries Investigated by Differential Voltage Analysis and Coulomb Tracking. *J. Electrochem. Soc.* **2017**, *164*, A6066. [CrossRef]

44. Xiao, J.; Li, Q.; Bi, Y.; Cai, M.; Dunn, B.; Glossmann, T.; Liu, J.; Osaka, T.; Sugiura, R.; Wu, B.; et al. Understanding and Applying Coulombic Efficiency in Lithium Metal Batteries. *Nat. Energy* **2020**, *5*, 561–568. [CrossRef]

45. Burns, J.C.; Stevens, D.A.; Dahn, J.R. In-Situ Detection of Lithium Plating Using High Precision Coulometry. *J. Electrochem. Soc.* **2015**, *162*, A959–A964. [CrossRef]

46. Adams, B.D.; Zheng, J.; Ren, X.; Xu, W.; Zhang, J.-G. Accurate Determination of Coulombic Efficiency for Lithium Metal Anodes and Lithium Metal Batteries. *Adv. Energy Mater.* **2018**, *8*, 1702097. [CrossRef]

47. Fang, C.; Li, J.; Zhang, M.; Zhang, Y.; Yang, F.; Lee, J.Z.; Lee, M.-H.; Alvarado, J.; Schroeder, M.A.; Yang, Y.; et al. Quantifying Inactive Lithium in Lithium Metal Batteries. *Nature* **2019**, *572*, 511–515. [CrossRef]

48. Iurilli, P.; Brivio, C.; Wood, V. On the Use of Electrochemical Impedance Spectroscopy to Characterize and Model the Aging Phenomena of Lithium-Ion Batteries: A Critical Review. *J. Power Source* **2021**, *505*, 229860. [CrossRef]

49. Westerhoff, U.; Kurbach, K.; Lienesch, F.; Kurrat, M. Analysis of Lithium-Ion Battery Models Based on Electrochemical Impedance Spectroscopy. *Energy Technol.* **2016**, *4*, 1620–1630. [CrossRef]

50. Bieker, G.; Winter, M.; Bieker, P. Electrochemical in Situ Investigations of SEI and Dendrite Formation on the Lithium Metal Anode. *Phys. Chem. Chem. Phys.* **2015**, *17*, 8670–8679. [CrossRef]

51. Tröltzsch, U.; Kanoun, O.; Tränkler, H.-R. Characterizing Aging Effects of Lithium Ion Batteries by Impedance Spectroscopy. *Electrochim. Acta* **2005**, *51*, 1664–1672. [CrossRef]

52. Zhang, Q.; Wang, D.; Schaltz, E.; Stroe, D.-I.; Gismero, A.; Yang, B. Degradation Mechanism Analysis and State-of-Health Estimation for Lithium-Ion Batteries Based on Distribution of Relaxation Times. *J. Energy Storage* **2022**, *55*, 105386. [CrossRef]

53. Iurilli, P.; Brivio, C.; Wood, V. Detection of Lithium-Ion Cells' Degradation through Deconvolution of Electrochemical Impedance Spectroscopy with Distribution of Relaxation Time. *Energy Technol.* **2022**, 2200547. [CrossRef]

54. Estaller, J.; Kersten, A.; Kuder, M.; Thiringer, T.; Eckerle, R.; Weyh, T. Overview of Battery Impedance Modeling Including Detailed State-of-the-Art Cylindrical 18650 Lithium-Ion Battery Cell Comparisons. *Energies* **2022**, *15*, 3822. [CrossRef]

55. Uddin, K.; Perera, S.; Widanage, W.D.; Somerville, L.; Marco, J. Characterising Lithium-Ion Battery Degradation through the Identification and Tracking of Electrochemical Battery Model Parameters. *Batteries* **2016**, *2*, 13. [CrossRef]

56. Product Specifications—Battery Tester BCS-800 Series. Available online: https://www.biologic.net/documents/hight-throughput-battery-tester-bcs-8xx-series/ (accessed on 16 April 2021).

57. Angelantoni Test Technologies Discovery Climatic Chambers. Available online: https://www.acstestchambers.com/en/environmental-test-chambers/discovery-my-climatic-chambers-for-stress-screening/ (accessed on 16 April 2021).

Article

Effects of Coating on the Electrochemical Performance of a Nickel-Rich Cathode Active Material

Eman Hassan [1,2], Mahdi Amiriyan [2], Dominic Frisone [1], Joshua Dunham [1,2], Rashid Farahati [2,*] and Siamak Farhad [1,*]

1   Advanced Energy and Manufacturing Laboratory, Department of Mechanical Engineering, University of Akron, Akron, OH 44325, USA; eh40@uakron.edu (E.H.); df99@uakron.edu (D.F.); dunhajsh@schaeffler.com (J.D.)

2   Schaeffler Group, Wooster, OH 44691, USA; amirimhd@schaeffler.com

*   Correspondence: farahrsh@schaeffler.com (R.F.); sfarhad@uakron.edu (S.F.)

Abstract: Due to their safety and high power density, one of the most promising types of all-solid-state lithium batteries is the one made with the argyrodite solid electrolyte (ASE). Although substantial efforts have been made toward the commercialization of this battery, it is still challenged by some technical issues. One of these issues is to prevent the side reactions at the interface of the ASE and the cathode active material (CAM). A solution to address this issue is to coat the CAM particles with a material that is compatible with both ASE and CAM. Prior studies show that the lithium niobate, $LiNbO_3$, (LNO) is a promising material for coating CAM particles to reduce the interfacial side reactions. However, no systematic study is available in the literature to show the effect of coating LNO on CAM performance. This paper aims to quantify the effect of LNO coating on the electrochemical performance of a nickel-rich CAM. The electrochemical performance parameters that are studied are the capacity, cycling performance, and rate performance of the coated-CAM; and the effectiveness of the coating to prevent the side reactions at the ASE and CAM interface is out of the scope of this study. To eliminate the effect of side reactions at the ASE and CAM interface, we conduct all tests in the organic liquid electrolyte (OLE) cells to solely present the effect of coating on the CAM performance. For this purpose, 0.5 wt.% and 1 wt.% LNO are used to coat the $LiNi_{0.6}Mn_{0.2}Co_{0.2}O_2$ (NMC-60) CAM through two synthesizing methods. Consequently, the effects of the synthesizing method and the coating weight percentage on the NMC-60 performance are presented.

Keywords: coating; wet process; nickel-rich NMC; lithium niobate; electrochemical performance

Citation: Hassan, E.; Amiriyan, M.; Frisone, D.; Dunham, J.; Farahati, R.; Farhad, S. Effects of Coating on the Electrochemical Performance of a Nickel-Rich Cathode Active Material. *Energies* **2022**, *15*, 4886. https://doi.org/10.3390/en15134886

Academic Editor: Cai Shen

Received: 8 June 2022
Accepted: 28 June 2022
Published: 3 July 2022

Publisher's Note: MDPI stays neutral with regard to jurisdictional claims in published maps and institutional affiliations.

1. Introduction

In traditional commercial lithium-ion batteries (LIBs), the use of liquid electrolytes containing flammable organic solvents creates potential safety issues [1]. All-solid-state lithium batteries (ASSLB), on the other hand, utilize intrinsically safe solid-state electrolytes [2,3]. Hence, they are considered safe next-generation battery systems, especially for applications in electric vehicles (EVs). There are several types of ASSLBs, depending on the type of solid-state electrolyte used to make the battery. Among all types of solid-state electrolytes, the argyrodite electrolyte (ASE), for example, $Li_6PS_5Cl_{0.5}Br_{0.5}$ (LPSCB), is considered one of the most promising electrolytes due to its high ionic conductivity and special mechanical properties [4]. The high ionic conductivity makes this solid electrolyte suitable for applications where high power density is required, for example, in hybrid electric vehicles (HEVs).

Significant efforts have been devoted toward the development of ASE-type ASSLBs by researchers. However, some technical issues still need to be addressed before the commercialization of these batteries. Some of these issues are (1) the interfacial resistance between the ASE and the cathode active material (CAM) in the cathode that causes an increase in the cathode ohmic resistance, consequently, the battery capacity decreases and the heat generation in the battery increases, (2) the side-reactions at the interface of the ASE and CAM that causes fast battery degradation, and (3) the scalability for large-scale manufacturing [4,5]. As proposed by Divakaran et al. [6] and other researchers, one practical method to suppress side reactions at the interface of the ASE, such as LPSCB, and CAM, such as lithium nickel manganese cobalt oxide (NMC) or lithium nickel cobalt aluminum oxide (NCA), is to coat them with a material that is not only ionically conductive and acts as a second electrolyte, but also it is compatible with ASE, CAM, conductive material, and cathode current collector [6,7]. The interface between the coating material and each of these materials is schematically shown in Figure 1.

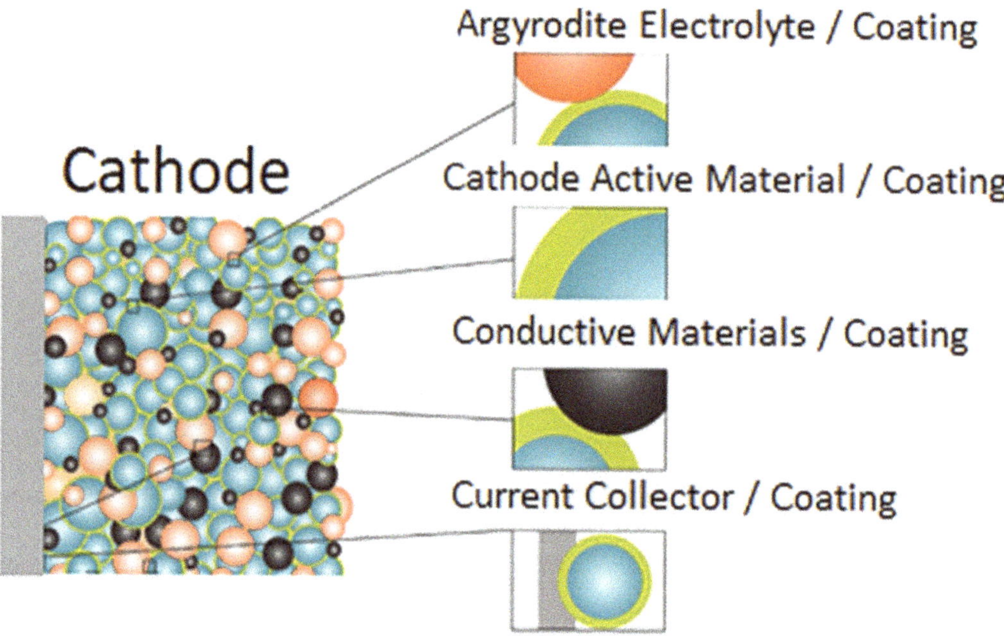

Figure 1. Interface of the coating material with other components of the cathode. The red particles are argyrodite electrolyte, the blue particles are the cathode active material, the black particles are the electron conductive materials, the gray layer is the cathode current collector, and the green layers are the coating (Reprinted/adapted with permission from Ref. [7]. Copyright year: 2022, copyright owner's name: Eman Hassan).

Several coating materials and procedures have been developed by researchers to suppress the interfacial side reactions between the electrolyte and electrode active materials. However, the addition of a surface coating layer can negatively impact ionic conductivity and decrease the cathode capacity and the rate performance [8]. Accordingly, researchers have reported compounds containing lithium as viable candidates for the coating of nickel-rich metal oxides because of their low impedance, increased ionic conductivity, and increased chemical stability [9,10]. Some compounds that have shown promise in resolving these issues are Li_2ZrO_3, $LiAlO_2$, Li_2TiO_3, and $LiNbO_3$ [11–14]. Lithium niobate, $LiNbO_3$, (LNO) has been suggested as a suitable coating material in several studies [3,14,15]. Due to its low detriment to conductivity [16,17], LNO has recently proven to be a viable transition metal oxide contender for coating as it increases Li^+ mobility at the cathode surface. Additionally, due to its high thermal stability, LNO allows for operation at high temperatures for long periods of time without negative levels of dissolution. Despite the above-mentioned benefits, the effects of coating this material, along with its synthesis methods and coating thicknesses, on the performance of CAM still need to be systematically investigated.

Various methods can be employed to deposit coatings on the surface of cathode active material particles. Examples of methods that have been utilized by researchers for this purpose are dry coating, atomic layer deposition (ALD), and wet mixing [18–21]. Among these processes to coat NMC with LNO, the application of a relatively simple wet process followed by heat treatment has shown merits, while ALD is also promising. The procedure for this method is simple enough that issues of scalability for mass manufacturing can be resolved. The initial step of the wet process method involves the dissolution of LNO precursors in a solvent. Mereacre et al. [22] have shown that along with this solvent the addition of hydrogen peroxide might improve the LNO coating through surface activation [22]. Li et al. [20] investigated the effect of LNO coating on nickel-rich NMC using a wet mixing method followed by heat treatment. Their study into NMC structure and particle morphology showed that LNO-coated NMC was very stable and presented a uniform coating on NMC particles. By evaluating the effects of LNO coating on the chemical, structural, and thermal stability of nickel-rich NMC, it was proven that LNO coating can improve the electrochemical performance of the cathode, especially at elevated temperatures [20]. Furthermore, there is much flexibility regarding heat treatment of the coated NMC. As shown by Kim et al. [21], the development of the LNO surface coating is contingent upon the sintering temperature. Kim et al. [21] sought several benefits from coating NMC with LNO, which were namely chemical, structural, and thermal stability. For their study, when LNO coating was heated at 450 °C, it was amorphously present on the surface of NMC. However, it showed crystallinity when heated at 800 °C. They found that desirable properties were provided by both the crystalline and amorphous structures. However, these valuable properties were found to a larger degree in the crystalline coating than in the amorphous coating [21].

There is not any systematic study in the literature to show the effect of coating LNO material on the CAM performance. This paper aims to quantify the effect of LNO coating on the electrochemical performance of the nickel-rich $LiNi_{0.6}Mn_{0.2}Co_{0.2}O_2$ (NMC-60) cathode active material. The electrochemical performance parameters that are studied are the capacity, cycling performance, and rate performance of the LNO-coated NMC-60. It is noted that the study on the effectiveness of the coating to prevent side reactions at the ASE and CAM interface is out of the scope of this paper. To eliminate the effect of the ASE and CAM side reactions, we conduct all tests in the organic liquid electrolyte (OLE) environment to solely study the effect of coating on the NMC-60 material performance. It is also noted that the reason for coating the CAM is to use them in the ASE-type ASSLBs, rather than increasing the CAM performance to use it in conventional OLE-type LIBs. The only reason that we choose testing cells in the OLE environment is to separate the effect of side reactions at the interface of the ASE and CAM from the effect of the LNO coating layer on the NMC-60 material performance. For this purpose, we (a) compare two methods of synthesizing and coating LNO on nickel-rich NMC-60 cathode active material, and (b) evaluate the effects of coating thickness on the capacity, cycling performance, and rate performance of the coated-NMC-60. This study not only helps to fabricate high-performance solid-state lithium batteries, but it also aids several other studies such as modeling solid-state lithium batteries for investigation of cell operating voltage and capacity [23], microstructure heterogeneity [24], battery energy efficiency [25], and designing an appropriate cooling system for the battery [26].

2. Materials and Methods

To synthesize the LNO coating, the following chemical reaction was utilized.

$$1\ CH_3CH_2OLi + 1\ Nb(CH_3CH_2O)_5 + 3H_2O \rightarrow 1\ LiNbO_3 + 6\ CH_3CH_2OH$$

The synthesizing and coating methods of LNO that we used in this study are briefly explained below.

Method-I: For coating NMC-60 using Method-I, appropriate amounts of lithium ethoxide (CH_3CH_2OLi, 95%, Sigma-Aldrich, St. Louis, MO, USA) and niobium ethoxide

(Nb(CH$_3$CH$_2$O)$_5$, 99%, Sigma-Aldrich) are dissolved in dry isopropanol (99%, Sigma-Aldrich) and continuously stirred at room temperature for 10 min. Then, NMC-60 is added to the solution and mixed at 80 °C until evaporation.

Method-II: In this method, lithium ethoxide (CH$_3$CH$_2$OLi, 95%, Sigma-Aldrich) and niobium ethoxide (Nb(CH$_3$CH$_2$O)$_5$, 99%, Sigma-Aldrich) are mixed in dry ethanol (99%, Sigma-Aldrich) and hydrogen peroxide (H$_2$O$_2$, 30%, Sigma-Aldrich). After dissolution of precursors under stirring for 10 min, NMC-60 is added to the solution and mixed at 100 °C until evaporation.

For both methods, the resulting dried powders are placed in zirconia combustion boats (AdValue Technology) and sealed in a quartz tube furnace (GSL-1100X, MTI Corporation). Powders are then annealed under flowing O$_2$ atmosphere at 450 °C for 1 h with a ramp of 5 °C/min. The coated powders are collected, ground, and kept overnight at 100 °C in a vacuum oven before making electrodes.

3. Coating Formulations

The entire mass of coating material does not participate in forming a solid and dense coating layer on the surface of CAM particles and remains as loosely connected LNO to CAM particles or agglomerated LNO as impurities in the obtained coated-CAM. Hence, we define the coating efficiency as the ratio of the mass of the dense coating layer to the total mass of the coating material as stated in Equation (1).

$$\eta_{coating} = \frac{m_{DCL}}{m_{CAM}} \tag{1}$$

where, $\eta_{coating}$ is the coating efficiency and m is the mass. The subscripts of DCL and CAM denote the dense coating layer and cathode active material, respectively. For one of the samples of 1 wt.% of LNO coating on NMC-60, we did the transmission electron microscopy (TEM) characterization at Argonne National Laboratory, Center for Nanoscale Materials (CNM). Based on the TEM image in Figure 2, the thickness of the dense coating layer on the NMC-60 is about 14 nm, while the 1 wt.% LNO is enough to make a 21 nm coating layer. Therefore, about two-thirds of the theoretical LNO thickness is formed as a dense coating layer. The other one-third of LNO material remains as loosely connected to NMC-60, or agglomerated LNO as impurities in the obtained coated-NMC-60. If required, the loosely connected LNO and the LNO impurities may be removed by rinsing the coated-NMC after the coating process is done. Therefore, we assume a coating efficiency of 66% for the calculation of the LNO coating layer thickness, which may be only valid for our lab with its available equipment and technology. It is also noted that the thickness of the LNO coating on the NMC-60 may not be uniform. This may affect the value of the coating efficiency.

The mass fraction ratio (MFR) of the coating material (LNO in this study) and the uncoated active material (NMC-60 in this study) is defined in Equation (2).

$$
\begin{aligned}
MFR &= \frac{MF_{CM}}{MF_{CAM}} = \frac{m_{CM}/m_{tot}}{m_{CAM}/m_{tot}} = \frac{m_{CM}}{m_{CAM}} = \frac{1}{\eta_{coating}} \times \frac{m_{DCL}}{m_{CAM}} = \frac{1}{\eta_{coating}} \times \frac{\sum \rho_{CM} N_{p,i} V_{DCL,i}}{\sum \rho_{CAM} N_{p,i} V_{am,i}} \\
&= \frac{1}{\eta_{coating}} \times \frac{\rho_{CM}}{\rho_{CAM}} \times \frac{\sum N_{p,i} \pi D_{CAM,i}^2 \delta_{DCL}}{\sum N_{p,i} \frac{\pi}{6} D_{CAM,i}^3} \\
&= \frac{1}{\eta_{coating}} \times \left(6\frac{\rho_{CM}}{\rho_{CAM}} \times \frac{\sum N_{p,i} D_{CAM,i}^2}{\sum N_{p,i} D_{CAM,i}^3} \right) \delta_{DCL} \\
&= \frac{1}{\eta_{coating}} \times \left(6\frac{\rho_{CM}}{\rho_{CAM}} \times \frac{D_{s,CAM}^2}{D_{v,CAM}^3} \right) \delta_{DCL}
\end{aligned}
\tag{2}
$$

where, MF is the mass fraction, ρ is the density, N_p is the number of CAM particles, V is the volume, and δ_{DCL} is the dense coating thickness on the active material. The subscripts of s and v denote the diameter of uncoated CAM, the surface mean diameter of particles, and the volume mean diameter of particles, respectively. The density of the LNO and NMC are approximately 4.65 g/cm^3 and 4.76 g/cm^3, respectively.

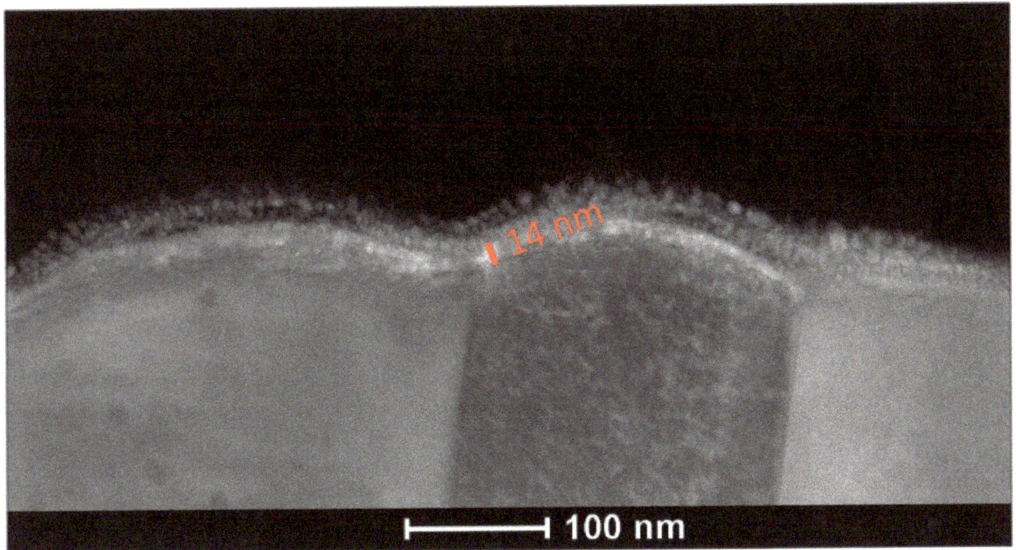

Figure 2. TEM image of the LNO coated NMC-60 showing a dense coating layer with a thickness of 14 nm has been formed, which is two-thirds of the theoretical thickness of 21 nm for a 1 wt.% of LNO coating on NMC-60 (work performed at the Center for Nanoscale Materials at Argonne National Laboratory).

To calculate the *MFR*, we need to determine the surface mean and volume mean diameters of uncoated CAM particles. For this purpose, the morphology of the NMC-60 was determined using Tescan Lyra 3 XMU scanning electron microscopy (SEM) at an operating voltage of 15 kV with an EDAX Element energy-dispersive X-ray spectroscopy (EDX) detector. The SEM image of the uncoated NMC-60 is taken as shown in Figure 3a. As seen, the shape of particles is close to a sphere. Thus, to obtain the sizes of particles, we assumed that the particles are spherical, and using the *ImageJ* software, we measured the diameter of more than 300 particles of NMC-60. Then, the measured particles are divided into several intervals and the histogram of the particle size distribution is plotted. Based on the obtained histogram, it was determined that a log-normal distribution is the best fit to the size distribution of NMC-60 particles, as shown in Figure 3b.

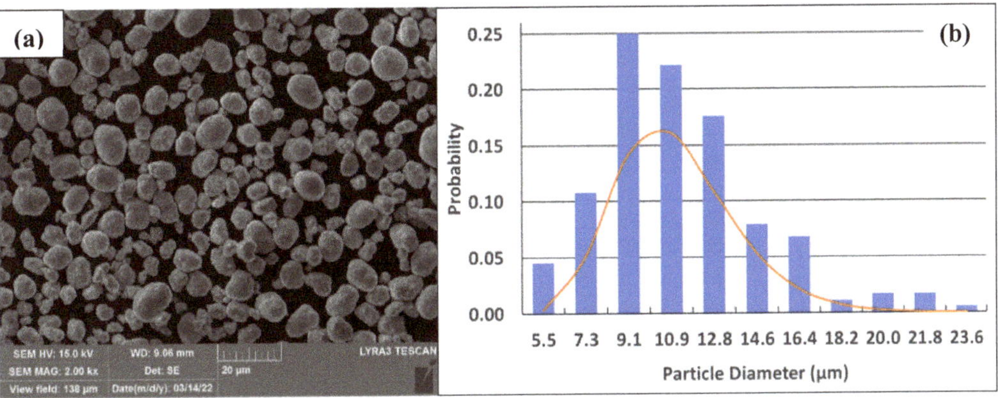

Figure 3. (**a**) SEM of the uncoated NMC-60 particles, and (**b**) approximation of NMC-60 particle sizes with log-normal distribution.

The surface mean and volume mean diameters for the log-normal distribution are determined from Equations (3) and (4), respectively. For details, the readers are referred to the authors' other publications in Ref. [27].

$$D_s = \left(\frac{\sum N_{p_i} D_i^2}{N_p} \right)^{1/2} = \frac{D_{50}}{exp\left(-\left(ln\left(\frac{D_{84.13}}{D_{50}} \right) \right)^2 \right)} \tag{3}$$

$$D_v = \left(\frac{\sum N_{p_i} D_i^3}{N_p} \right)^{1/3} = \frac{D_{50}}{\left[exp\left(-\left(ln\left(\frac{D_{84.13}}{D_{50}} \right) \right)^2 \right) \right]^{1.5}} \tag{4}$$

where, D_s is the surface mean diameter, D_v is the volume mean diameter, D_{50} is the median diameter, and $D_{84.15}$ is the diameter that 84.15% of particles are smaller than. The D_{50} and $D_{84.15}$ are determined by analyzing SEM images of NMC-60 particles using *ImageJ* software. For NMC-60, we obtained D_{50} and $D_{84.15}$ are 10.9 μm and 13.7 μm, respectively. Finally, we calculated that the D_s and D_v are 11.5 μm, and 11.8 μm, respectively.

For preparation of the LNO coated NMC-60 with the desired coating thickness, whether using Method-I or Method-II, the mass fraction of initial materials for CH_3CH_2OLi, $Nb(CH_3CH_2O)_5$, and uncoated NMC-60 are obtained from Equations (5)–(7).

$$MF_{CH_3CH_2OLi} = \frac{MW_{CH_3CH_2OLi}}{MW_{LNO}} \times \left(1 + \frac{1}{MFR} \right)^{-1} \tag{5}$$

$$MF_{Nb(CH_3CH_2O)_5} = \frac{MW_{Nb(CH_3CH_2O)_5}}{MW_{LNO}} \times \left(1 + \frac{1}{MFR} \right)^{-1} \tag{6}$$

$$MF_{NMC\_60} = \frac{1}{1 + MFR} \tag{7}$$

The molecular weights of CH_3CH_2OLi, $Nb(CH_3CH_2O)_5$, and LNO are 52.0 g/mol, 318.21 g/mol, and 147.85 g/mol, respectively.

4. Electrode and Cell Fabrication and Testing

Several OLE-type half-cell cathodes with mass loading of ~7 mg/cm^2 were made to evaluate and compare the performance of the coated and uncoated NMC-60. The cathodes were made from four types of coated NMC-60: 0.5% LNO coated NMC-60 using Method-I (Method-I-0.5%), 1% LNO coated NMC-60 using Method-I (Method-I-1%), 0.5% LNO coated NMC-60 using Method-II (Method-II-0.5%), and 1% LNO coated NMC-60 using Method-II (Method-II-1%). First, a 6 wt.% PVDF (Sigma-Aldrich) solution is made by dissolution in NMP (99.5%, Sigma-Aldrich) and allowed to mix. A cathode slurry comprising of 90 wt.% active material, 5 wt.% conductive material and 5 wt.% binder was then made. The appropriate amounts of acetylene black (MTI corporation) and the PVDF solution were mixed in a planetary centrifugal mixer. One-third of the coated active material is then mixed with the acetylene black and PVDF solution. This step is repeated until the total amount of active material has been added and mixed. Then, several coin type half cells were fabricated using the OLE (LiPF$_6$), coated NMC-60 as the cathode, and the lithium metal as anode (reference electrode) with a separator in between.

Before testing the half-cells, a formation process was completed. For the rate performance test, we cycled the cathode half-cells at c-rates of C/10, C/5, C/3, 1C, 2C, and C/10, with five cycles at each C-rate (CC C/25 CV Charge; Discharge: No CV mode). The C-rate is defined as the rate at which a cell completely discharges its maximum capacity. To test half-cells using NMC-60 as cathode, a theoretical maximum capacity of 178 mAhg^{-1} is assumed.

5. Results and Discussion

5.1. Coating Formation

The mass fractions of initial materials for coating NMC-60 were calculated from the desired coating thickness and shown in Figure 4. The red and orange lines in this figure represent the mass fractions of lithium ethoxide and niobium ethoxide in the synthesizing process, respectively. The green line indicates the mass fraction of the total LNO coating material synthesized, while the blue line indicates the mass fraction of LNO participated to form the dense coating layer on NMC-60 particles. The LNO that has not participated in formation of the dense coating layer remains as loosely connected LNO to NMC-60 particles, or agglomerated LNO as impurities in the obtained coated- NMC-60, as demonstrated in red circles in Figure 5.

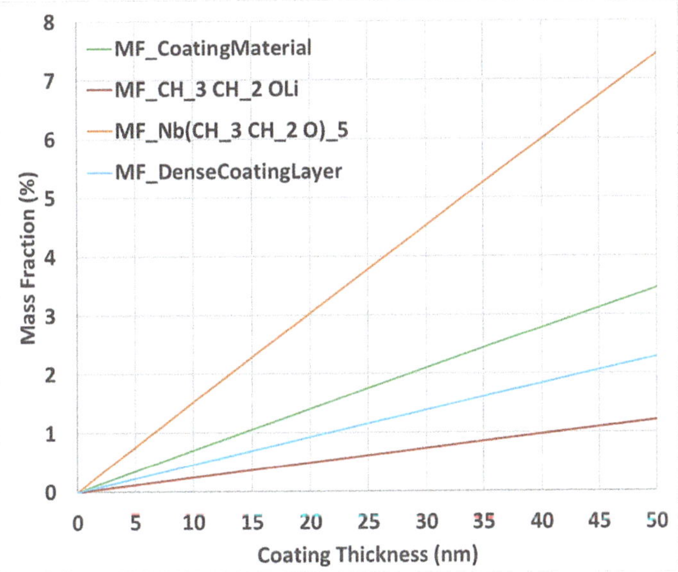

Figure 4. Mass fraction of initial materials to coat NMC-60 with the desired thickness of LNO.

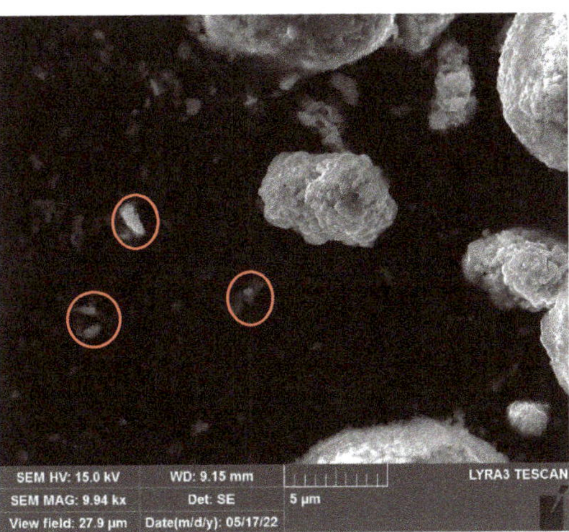

Figure 5. Materials in red circles seem to be the agglomerated LNO with sizes < 1 µm produced during synthesizing, but has not participated in coating layer formation on the NMC-60 particles.

The SEM images was used to evaluate the morphology of LNO coating on NMC-60 using Method-I and Method-II. A comparison of changes in morphology between uncoated NMC-60, Method-I-1%, and Method-II-1% can be seen through the SEM images in Figure 6a–c, respectively. As shown in Figure 6a, the uncoated NMC-60 appears to have particles that are mostly spherical with some irregular edges. On the other hand, images of the coated NMC-60 show, in appearance, a white material sitting on the surface of particles. This is most clearly seen on the large particles of Method-I coated NMC-60 in Figure 6b. Method-I also showed agglomeration of secondary particles, which is not present in uncoated NMC-60 nor Method-II-1%. It is essential to note how the primary and secondary particles become less defined for the coated powders when compared to uncoated NMC-60. For instance, in the Method-II coating, the white material appearing in the images of the coated NMC-60 takes on a feathery appearance, which covers the entirety of some NMC particles. An enlarged image of a particle which exhibits this phenomenon is inset in Figure 6c. Figure 7 shows the corresponding EDX analysis for SEM imaging of the coated particles. Formation of LNO coating on NMC-60 particles is indicated through the presence of niobium elemental peaks for Method-I and Method-II, respectively.

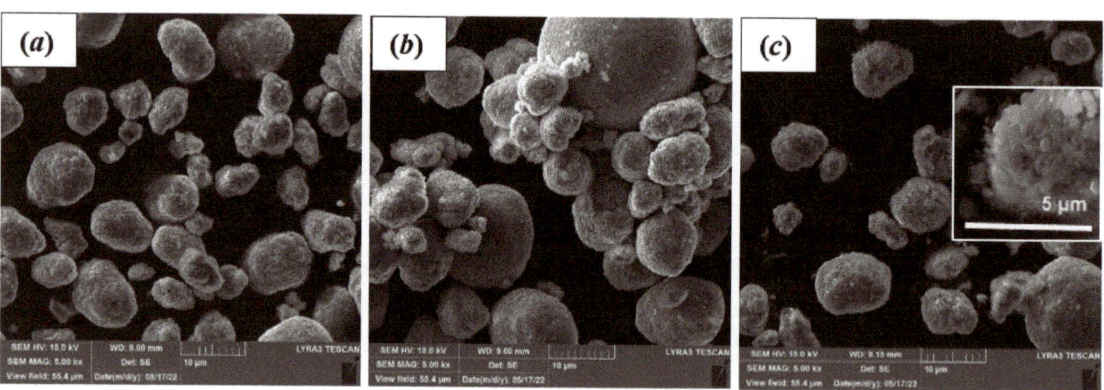

Figure 6. SEM imaging of (**a**) uncoated NMC-60, (**b**) Method-I-1%, and (**c**) Method-II-1% with inset exhibiting LNO as soft or feathery surface material in appearance.

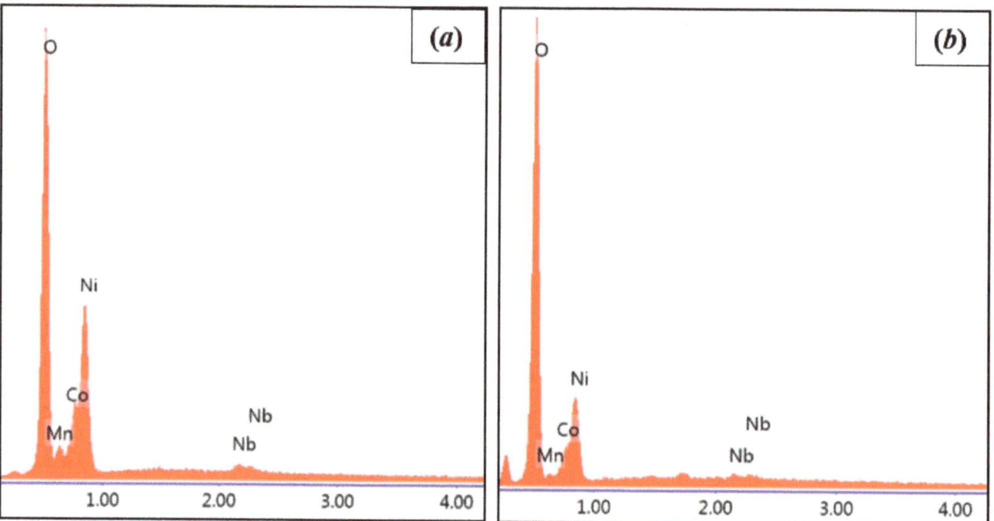

Figure 7. EDX analysis of 1% LNO coating using (**a**) Method-I and (**b**) Method-II.

The powder XRD measurements were taken using Rietveld analysis to compare the effect of LNO coating on NMC-60. Patterns of uncoated and 1% LNO coated NMC-60 using Method-I and Method-II are shown in Figure 8a–c. The XRD patterns of the three materials can be indexed to a hexagonal α-NaFeO$_2$ structure [20]. Moreover, there are no extra peaks on the coated NMC-60 materials indicating that no structural changes occurred as a result of the LNO coating or annealing procedure. It is noted that the LNO peaks are not visible in the XRD patterns due to the very small composition in the overall material being analyzed.

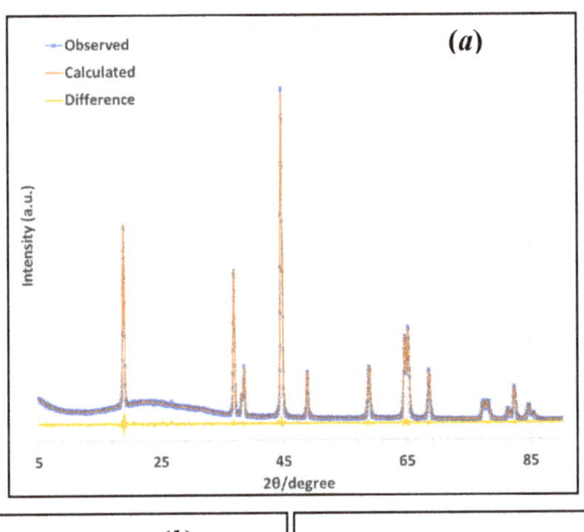

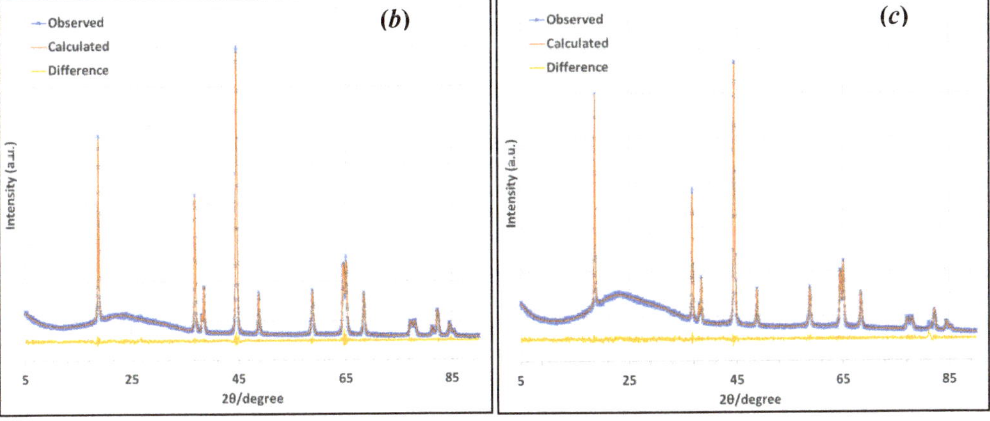

Figure 8. XRD analysis of (**a**) uncoated NMC-60, (**b**) 1% LNO coated NMC-60 using Method-I, and (**c**) 1% LNO coated NMC-60 using Method-II.

5.2. Electrochemical Performance

Electrochemical performance study of uncoated NMC-60, coated using Method-I, and coated using Method-II, with 0.5 wt.% and 1 wt.% LNO content for both methods, was performed in the voltage range of 2.7 V to 4.3 V using coin half-cells. Two tests were performed. The cycling performance was the first test which was used to evaluate the stability and capacity retention of the cathode half-cells over time. The CAM capacity was also determined during this test. The rate performance was the second test used to evaluate the power performance under different current loadings.

5.2.1. Cycling Performance Test

Cycling performance was performed at a charge and discharge rate of C/3 for 50 cycles. The results of this performance for uncoated NMC-60, Method-I, and Method-II with 0.5% and 1% for both methods are shown in Figure 9. Each point in this graph was found through charge and discharge voltage versus capacity curves. These curves showed almost the same trend between uncoated and coated NMC-60 with only changes in values of the capacity. Half-cells made using uncoated NMC-60 show an initial capacity of ~164.0 mAhg^{-1} and a capacity retention of 98.0%. Method-I-1% showed comparative values with an initial capacity of 166.4 mAhg^{-1} and a capacity retention of 97.0%. In comparison, Method-I-0.5% has a higher initial capacity of ~167.0 mAhg^{-1} with a much higher capacity retention of ~99.9%. This capacity retention is higher than the uncoated NMC-60, may be due to the protective coating layer formed around the NMC-60 particles. On the other hand, Method-II-1% showed an initial capacity of ~155.1 mAhg^{-1}, which is much lower than that of the uncoated NMC-60 or Method-I-1%. However, the capacity retentions of uncoated NMC-60 (98.0%) and Method-II-1% (97.7%) were very similar. Method-II-0.5% has an initial capacity of 171.0 mAhg^{-1} which is higher than all other initial capacities. Contrastingly, while Method-II-0.5% has higher initial capacity, capacity retention is much greater for Method-I-0.5% with a value of ~99.9% versus 97.0% for Method-II-0.5%. It is apparent that Method-I-0.5% has very good stability with almost no capacity fade after 50 cycles. Furthermore, with special regard to capacity fade, Method-I-0.5% shows a clear improvement in electrochemical performance of NMC-60 with organic liquid electrolytes. This improvement should be checked with the argyrodite electrolyte as well, which is the out of the scope of this paper.

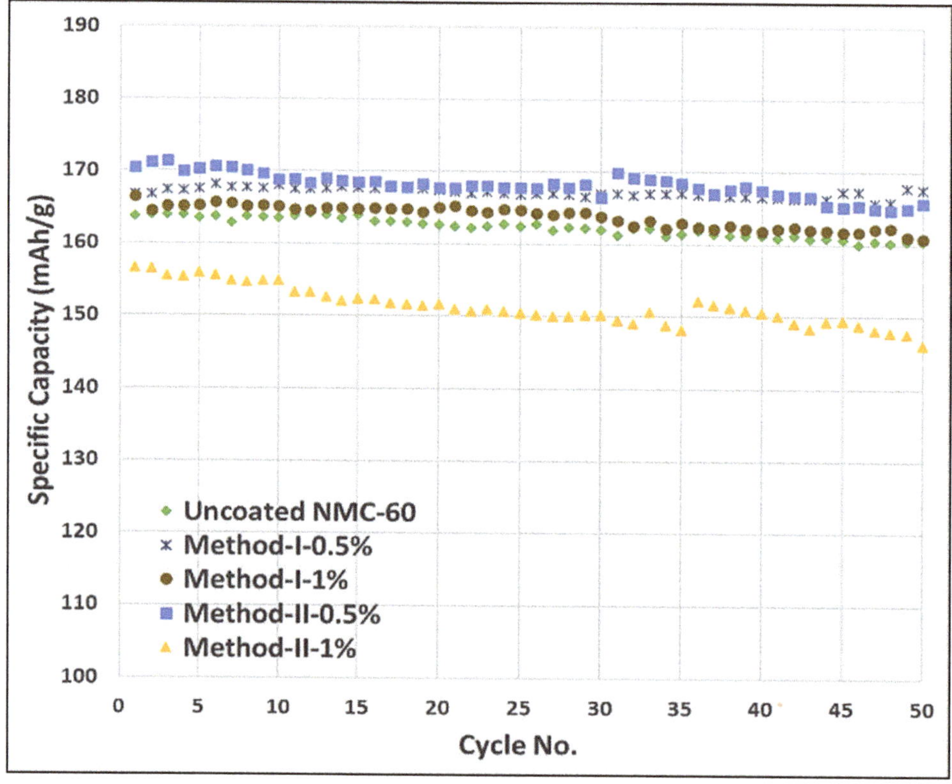

Figure 9. Cycling performance of the uncoated NMC-60 and 0.5 wt.% and 1 wt.% LNO coating NMC-60 using Method-I and Method-II.

We define the capacity retention ratio (CRR) as the ratio of the capacity retention of the coated-CAM to the capacity retention of the uncoated-CAM. The CRR of the coated-NMC-60 with methods I and II is shown in Figure 10. In this figure, the CRR of the uncoated NMC-60 has been represented by the LNO wt.% of 0, which is obviously equal to 1. The Method-I-0.5% exhibits CRR $\approx$ 1.026 after 50 cycles, which is higher than other coated samples. The CRR of Method-I-1% and Method-II-0.5% are less than that of the uncoated-NMC-60, while Method-II-1% shows almost the same CRR compared to the uncoated-NMC-60.

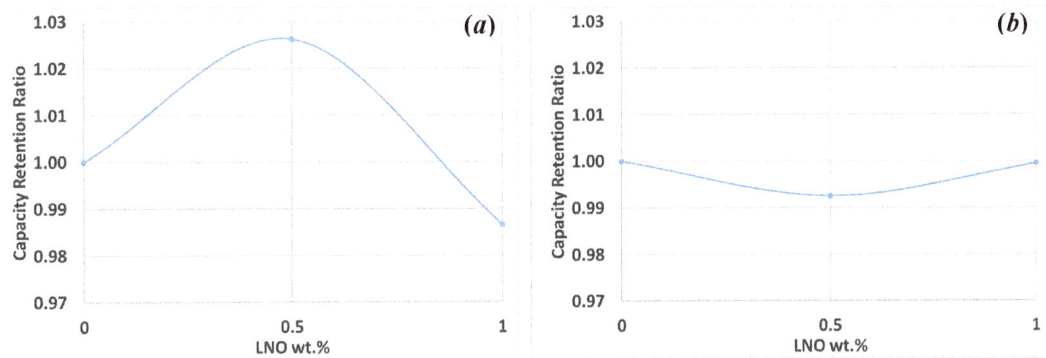

Figure 10. Capacity retention ratio of the coated-NMC-60 for (**a**) Method-I, and (**b**) Method-II.

5.2.2. Rate Performance Test

Results of the rate performance test to compare the uncoated NMC-60 with the coated NMC-60 with Method-I and Method-II 0.5 wt.% and 1 wt.% LNO are shown in Figure 11. Tests were run in the same cut-off voltage range of 2.7 V to 4.3 V. Different C-rates of C/10, C/5, C/3, C/2, C, and 2C followed by a return to C/10 were each tested for 5 cycles. Regarding initial capacity, Method-I-1% has higher values than the uncoated NMC-60 and Method-I-0.5%. Method-I-1% also exhibits these higher capacities for all c-rates. Method-II-0.5% has higher capacity than all other samples for all c-rates. Conversely, Method-II-1% has lower capacity than all other samples for all c-rates. Method-I-0.5% consistently performs at lower capacities than uncoated NMC-60 for c-rates below 2C. Although Method-II-0.5% has the highest capacity overall, this sample has the lowest recuperation of capacity after returning to C/10, while Method-I-0.5% has the highest. For better comparison of rate capabilities, Table 1 features the initial capacities, capacities at 2C, and capacities after returning to C/10 for all five samples as well as their corresponding capacity retentions.

It is evident from the data that the most improvement of NMC-60 regarding rate capability is achieved through Method-II-0.5% and Method-I-1%. These two samples exhibit higher capacities and capacity retention than the uncoated NMC-60 even at high c-rates. This trend is maintained upon returning to C/10 discharge rate. It should also be noted that Method-I-0.5% performs better than the uncoated NMC-60 after returning to C/10.

We define the specific capacity ratio (SCR) as the ratio of the specific capacity of the coated-CAM to the specific capacity of the uncoated-CAM. The SCR of the coated-NMC-60 with methods I and II is shown in Figure 12. In this figure, the SCR of the uncoated NMC-60 has been represented by the LNO wt.% of 0, which is obviously equal to 1. The comparison is presented for low versus high c-rates. It is evident from Figure 12a that the 1 wt.% LNO for Method-I can keep the specific capacity of the coated-NMC-60 about 1 to 2 percent more than the specific capacity of the uncoated-NMC-60 at both low and high C-rates of C/10 and 2C. On the other hand, Figure 12b shows that the 0.5 wt.% LNO for Method-II can keep the specific capacity of the coated-NMC-60 about 6 percent more than the specific capacity of the uncoated-NMC-60 at both low and high C-rates of C/10 and 2C. Further investigations into the effects of various LNO wt.% are necessary to conclude a precise wt.% of LNO for each method.

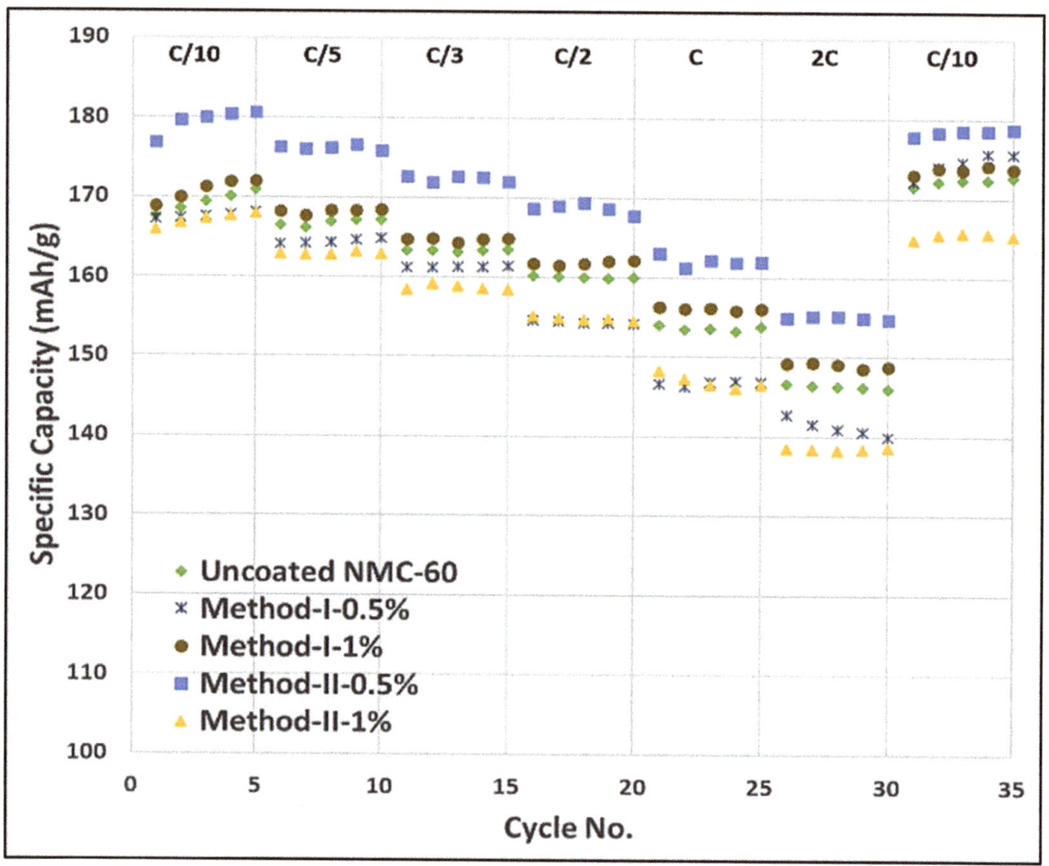

Figure 11. Rate performance of the uncoated NMC-60 and 0.5 wt.% and 1 wt.% LNO coating NMC-60 using Method-I and Method-II.

Table 1. Comparison of rate performance at low and high C-rates.

	Initial Capacity (mAhg^{-1})	Capacity at 2C (mAhg^{-1})	Capacity Retention at 2C (%)	Capacity upon Return to C/10 (mAhg^{-1})	Capacity Retention at C/10 (%)
Uncoated NMC-60	167.6	146.1	87.2	172.6	103.1
Method-I-0.5%	167.3	140.0	83.7	175.6	105.0
Method-I-1%	168.9	148.7	88.0	174.1	103.1
Method-II-0.5%	176.9	154.8	87.5	178.8	101.1
Method-II-1%	159.1	135.4	85.1	163.2	102.6

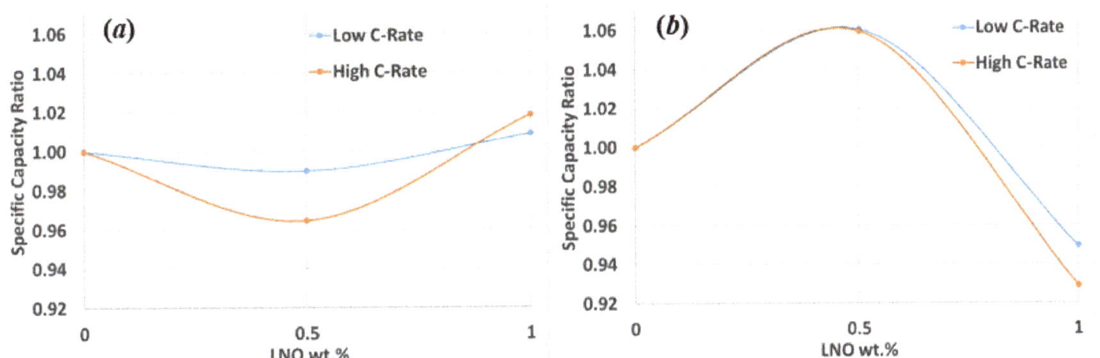

Figure 12. Specific capacity ratio of the coated-NMC-60 for (**a**) Method-I, and (**b**) Method-II for low or high c-rates of C/10 and 2C.

6. Conclusions

This paper aimed to quantify the effect of LNO coating on the electrochemical performance of the nickel-rich NMC-60 cathode active material. The electrochemical performance parameters of the initial capacity, cycling performance, and rate performance were studied. To eliminate the effect of side reactions at the interface of the argyrodite solid electrolyte and NMC-60, we conduct all tests using organic liquid electrolyte cells to solely study the effect of coating on the NMC-60 electrochemical performance. We presented a model and formulation to control the coating thickness on the electrode active material particles. Based on these formulations and by using two synthesizing and coating methods, several coated-NMC-60 with 0.5 wt.% and 1 wt.% LNO were prepared. The effects of LNO on the morphology and electrochemical performance of the coated-NMC-60 were investigated. Using the TEM and SEM images and EDS analysis the presence of LNO coating on the surface of the NMC-60 was determined. Further characterization using XRD showed that the coating methods did not change the structure of NMC-60. The electrochemical performance analysis results indicated that the capacity, cycling performance, and the rate performance of the LNO coated-NMC60 are sensitive to the LNO coating thickness (or wt.%) and the synthesizing and coating method. It was found that the initial capacity and rate performance of the 0.5 wt.% LNO-coated-NMC-60 using Method II are noticeably higher than those of the uncoated-NMC-60. The initial capacity and rate performance of the 1 wt.% LNO-coated-NMC-60 using Method I are only slightly higher than the uncoated-NMC-60. The initial capacity and rate performance of the 0.5 wt.% LNO-coated-NMC-60 using Method I and 1 wt.% LNO-coated-NMC-60 using Method II are lower than the uncoated-NMC-60. Although the 0.5 wt.% LNO-coated-NMC-60 using Method II is promising, a more detailed study is required to determine the optimum LNO wt.% and the best synthesizing and coating methods. Although this study provided a baseline for electrochemical performance of the coated-NMC-60, further investigations are required by testing the coated-NMC-60 in solid state cells.

Author Contributions: Conceptualization, M.A. and S.F.; Data curation, E.H., M.A. and D.F.; Formal analysis, E.H., M.A. and S.F.; Funding acquisition, R.F. and S.F.; Investigation, E.H., M.A., D.F., J.D., R.F. and S.F.; Methodology, M.A. and S.F.; Project administration, M.A.; Resources, R.F.; Supervision, R.F.; Visualization, J.D.; Writing—original draft, E.H.; Writing—review & editing, M.A., R.F. and S.F. All authors have read and agreed to the published version of the manuscript.

Funding: The funding was provided internally by the Schaeffler Company and the University of Akron.

Informed Consent Statement: Not applicable.

Acknowledgments: The support of the Schaeffler Company and The University of Akron is highly appreciated. Work performed at the Center for Nanoscale Materials, a U.S. Department of Energy Office of Science User Facility, was supported by the U.S. DOE, Office of Basic Energy Sciences, under Contract No. DE-AC02-06CH11357.

Conflicts of Interest: The authors declare no conflict of interest.

References

1. Goodenough, J.B.; Kim, Y. Challenges for Rechargeable Li Batteries. Chemistry of Materials. *Am. Chem. Soc.* **2010**, *22*, 587–603. [CrossRef]
2. Hao, F.; Han, F.; Liang, Y.; Wang, C.; Yao, Y. Architectural Design and Fabrication Approaches for Solid-State Batteries. *MRS Bull.* **2018**, *43*, 775–781. [CrossRef]
3. Zubair, M.; Li, G.; Wang, B.; Wang, L.; Yu, H. Electrochemical Kinetics and Cycle Stability Improvement with Nb Doping for Lithium-Rich Layered Oxides. *ACS Appl. Energy Mater.* **2018**, *2*, 503–512. [CrossRef]
4. Zhu, X.; Wang, K.; Xu, Y.; Zhang, G.; Li, S.; Li, C.; Zhang, X.; Sun, X.; Ge, X.; Ma, Y. Strategies to Boost Ionic Conductivity and Interface Compatibility of Inorganic—Organic Solid Composite Electrolytes. *Energy Storage Mater.* **2021**, *36*, 291–308. [CrossRef]
5. Divakaran, A.M.; Hamilton, D.; Manjunatha, K.N.; Minakshi, M. Design, Development and Thermal Analysis of Reusable Li-Ion Battery Module for Future Mobile and Stationary Applications. *Energies* **2020**, *13*, 1477. [CrossRef]
6. Divakaran, A.M.; Minakshi, M.; Bahri, P.A.; Paul, S.; Kumari, P.; Divakaran, A.M.; Manjunatha, K.N. Rational Design on Materials for Developing next Generation Lithium-Ion Secondary Battery. *Prog. Solid State Chem.* **2021**, *62*, 100298. [CrossRef]
7. Xiao, Y.; Wang, Y.; Bo, S.H.; Kim, J.C.; Miara, L.J.; Ceder, G. Understanding Interface Stability in Solid-State Batteries. *Nat. Rev. Mater.* **2020**, *5*, 105–126. [CrossRef]
8. Shi, J.L.; Qi, R.; Zhang, X.D.; Wang, P.F.; Fu, W.G.; Yin, Y.X.; Xu, J.; Wan, L.J.; Guo, Y.G. High-Thermal- and Air-Stability Cathode Material with Concentration-Gradient Buffer for Li-Ion Batteries. *ACS Appl. Mater. Interfaces* **2017**, *9*, 42829–42835. [CrossRef]
9. Ryu, W.G.; Shin, H.S.; Park, M.S.; Kim, H.; Jung, K.N.; Lee, J.W. Mitigating Storage-Induced Degradation of Ni-Rich $LiNi_{0.8}Co_{0.1}Mn_{0.1}O_2$ Cathode Material by Surface Tuning with Phosphate. *Ceram. Int.* **2019**, *45*, 13942–13950. [CrossRef]
10. Cho, D.H.; Jo, C.H.; Cho, W.; Kim, Y.J.; Yashiro, H.; Sun, Y.K.; Myung, S.T. Effect of Residual Lithium Compounds on Layer Ni-Rich $Li[Ni_{0.7}Mn_{0.3}]O_2$. *J. Electrochem. Soc.* **2014**, *161*, A920–A926. [CrossRef]
11. Meng, K.; Wang, Z.; Guo, H.; Li, X.; Wang, D. Improving the Cycling Performance of $LiNi_{0.8}Co_{0.1}Mn_{0.1}O_2$ by Surface Coating with Li_2TiO_3. *Electrochim. Acta* **2016**, *211*, 822–831. [CrossRef]
12. Song, B.; Li, W.; Oh, S.M.; Manthiram, A. Long-Life Nickel-Rich Layered Oxide Cathodes with a Uniform Li_2ZrO_3 Surface Coating for Lithium-Ion Batteries. *ACS Appl. Mater. Interfaces* **2017**, *9*, 9718–9725. [CrossRef] [PubMed]
13. Srur-Lavi, O.; Miikkulainen, V.; Markovsky, B.; Grinblat, J.; Talianker, M.; Fleger, Y.; Cohen-Taguri, G.; Mor, A.; Tal-Yosef, Y.; Aurbach, D. Studies of the Electrochemical Behavior of $LiNi_{0.80}Co_{0.15}Al_{0.05}O_2$ Electrodes Coated with $LiAlO_2$. *J. Electrochem. Soc.* **2017**, *164*, A3266–A3275. [CrossRef]
14. Zhang, Z.J.; Chou, S.L.; Gu, Q.F.; Liu, H.K.; Li, H.J.; Ozawa, K.; Wang, J.Z. Enhancing the High Rate Capability and Cycling Stability of $LiMn_2O_4$ by Coating of Solid-State Electrolyte $LiNbO_3$. *ACS Appl. Mater. Interfaces* **2014**, *6*, 22155–22165. [CrossRef] [PubMed]
15. Xin, F.; Zhou, H.; Chen, X.; Zuba, M.; Chernova, N.; Zhou, G.; Whittingham, M.S. Li–Nb–O Coating/Substitution Enhances the Electrochemical Performance of the $LiNi_{0.8}Mn_{0.1}Co_{0.1}O_2$ (NMC 811) Cathode. *ACS Appl. Mater. Interfaces* **2019**, *11*, 34889–34894. [CrossRef]
16. Ohta, N.; Takada, K.; Sakaguchi, I.; Zhang, L.; Ma, R.; Fukuda, K.; Osada, M.; Sasaki, T. $LiNbO_3$-Coated $LiCoO_2$ as Cathode Material for All Solid-State Lithium Secondary Batteries. *Electrochem. Commun.* **2007**, *9*, 1486–1490. [CrossRef]
17. Gabrielli, G.; Axmann, P.; Diemant, T.; Behm, R.J.; Wohlfahrt-Mehrens, M. Combining Optimized Particle Morphology with a Niobium-Based Coating for Long Cycling-Life, High-Voltage Lithium-Ion Batteries. *ChemSusChem* **2016**, *9*, 1670–1679. [CrossRef]
18. Nakamura, H.; Kawaguchi, T.; Masuyama, T.; Sakuda, A.; Saito, T.; Kuratani, K.; Ohsaki, S.; Watano, S. Dry Coating of Active Material Particles with Sulfide Solid Electrolytes for an All-Solid-State Lithium Battery. *J. Power Sources* **2020**, *448*, 227579. [CrossRef]
19. Li, X.; Liu, J.; Banis, M.N.; Lushington, A.; Li, R.; Cai, M.; Sun, X. Atomic Layer Deposition of Solid-State Electrolyte Coated Cathode Materials with Superior High-Voltage Cycling Behavior for Lithium Ion Battery Application. *Energy Environ. Sci.* **2014**, *7*, 768–778. [CrossRef]
20. Li, X.; Jin, L.; Song, D.; Zhang, H.; Shi, X.; Wang, Z.; Zhang, L.; Zhu, L. $LiNbO_3$-Coated $LiNi_{0.8}Co0.1Mn_{0.1}O_2$ Cathode with High Discharge Capacity and Rate Performance for All-Solid-State Lithium Battery. *J. Energy Chem.* **2020**, *40*, 39–45. [CrossRef]
21. Kim, J.H.; Kim, H.; Choi, W.; Park, M.S. Bifunctional Surface Coating of $LiNbO_3$ on High-Ni Layered Cathode Materials for Lithium-Ion Batteries. *ACS Appl. Mater. Interfaces* **2020**, *12*, 35098–35104. [CrossRef] [PubMed]
22. Mereacre, V.; Stüble, P.; Ghamlouche, A.; Binder, J.R. Enhancing the Stability of $Lini_{0.5}Mn_{1.5}O_4$ by Coating with $LinbO_3$ Solid-State Electrolyte: Novel Chemically Activated Coating Process versus Sol-Gel Method. *Nanomaterials* **2021**, *11*, 548. [CrossRef] [PubMed]

23. Mastali, M.; Samadani, E.; Farhad, S.; Fraser, R.; Fowler, M. Three-Dimensional Multi-Particle Electrochemical Model of LiFePO$_4$ Cells Based on a Resistor Network Methodology. *Electrochim. Acta* **2016**, *190*, 574–587. [CrossRef]
24. Kashkooli, A.G.; Amirfazli, A.; Farhad, S.; Lee, D.U.; Felicelli, S.; Park, H.W.; Feng, K.; De Andrade, V.; Chen, Z. Representative Volume Element Model of Lithium-Ion Battery Electrodes Based on X-Ray Nano-Tomography. *J. Appl. Electrochem.* **2017**, *47*, 281–293. [CrossRef]
25. Farhad, S.; Nazari, A. Introducing the Energy Efficiency Map of Lithium-Ion Batteries. *Int. J. Energy Res.* **2019**, *43*, 931–944. [CrossRef]
26. Mohammed, A.H.; Esmaeeli, R.; Aliniagerdroudbari, H.; Alhadri, M.; Hashemi, S.R.; Nadkarni, G.; Farhad, S. Dual-Purpose Cooling Plate for Thermal Management of Prismatic Lithium-Ion Batteries during Normal Operation and Thermal Runaway. *Appl. Therm. Eng.* **2019**, *160*, 114106. [CrossRef]
27. Frisone, D.; Amiriyan, M.; Hassan, E.; Dunham, J.; Farahati, R.; Farhad, S. Effect of LiNbO$_3$ Coating on Capacity and Cycling of Nickle-Rich NMC Cathode Active Material. In Proceedings of the International Mechanical Engineering Congress and Exposition, Virtual, Online, 4 November 2021.

 *energies*

Review

A Review on the Molecular Modeling of Argyrodite Electrolytes for All-Solid-State Lithium Batteries

Oluwasegun M. Ayoola, Alper Buldum *, Siamak Farhad * and Sammy A. Ojo

Department of Mechanical Engineering, The University of Akron, Akron, OH 44325, USA
* Correspondence: buldum@uakron.edu (A.B.); sfarhad@uakron.edu (S.F.)

Abstract: Solid-state argyrodite electrolytes are promising candidate materials to produce safe all-solid-state lithium batteries (ASSLBs) due to their high ionic conductivity. These batteries can be used to power electric vehicles and portable consumer electronics which need high power density. Atomic-scale modeling with ab initio calculations became an invaluable tool to better understand the intrinsic properties and stability of these materials. It is also used to create new structures to tailor their properties. This review article presents some of the recent theoretical investigations based on atomic-scale modeling to study argyrodite electrolytes for ASSLBs. A comparison of the effectiveness of argyrodite materials used for ASSLBs and the underlying advantages and disadvantages of the argyrodite materials are also presented in this article.

Keywords: argyrodite electrolyte; all-solid-state lithium batteries; modeling; molecular dynamics; simulation

Citation: Ayoola, O.M.; Buldum, A.;
Farhad, S.; Ojo, S.A. A Review on the
Molecular Modeling of Argyrodite
Electrolytes for All-Solid-State
Lithium Batteries. *Energies* **2022**, *15*,
7288. https://doi.org/10.3390/
en15197288

Academic Editors: Antonino S. Aricò
and Francesco Lufrano

Received: 26 July 2022
Accepted: 28 September 2022
Published: 4 October 2022

Publisher's Note: MDPI stays neutral
with regard to jurisdictional claims in
published maps and institutional affiliations.

1. Introduction

Argyrodite electrolytes are very promising solid-state electrolytes (SSE) materials for all-solid-state batteries (ASSB). The high lithium mobility and conductivity in argyrodite electrolytes make it a potential solution for ASSBs [1]. The SSE also allows more effective cell packaging accompanied by their high gravimetric/volumetric energy density applications [2]. Lithium batteries with liquid organic electrolytes [3] have protection issues in terms of high combustibility and electrolyte leakages, while those with SSE are precisely contrary to these qualities [4–7]. Recently, Zhou et al. [8] have researched the electrochemical performance of Li_6PS_5Cl in an ideal solid-state battery (SSB). It exhibited sublime performance as its ionic conductivity is improved through the initiation of Li vacancies and disorder and showed congruence between cells. Some research works have been conducted to improve the activity of lithium-ion batteries, such as elastic stiffening tendency to increase stresses to aggrandize lithium-ion diffusion and uphold the elastic modulus [9].

Considering the SSEs, everything is not perfect. The electrolyte interfaces with both cathode and anode require significant considerations such as the restraint of the common boundary developed at the borderline of the cathode [10,11] and the anode [12,13]. Malleable solid electrolytes having an appropriate Young's modulus can repress the growth of lithium dendrite for lithium anode and enhance finer connection upon cycling when taken with the cathodic substance [3,14]. However, an established scalable route to such material is a substantial challenge that could limit the progress of the SSB field because of simple economics [8].

The investigation of the conductivity of SSE is an involved experiment because of the problem in the reproduced reformation of the materials and the responsivity of the conduction property. Computational approaches often utilize a measure for the determination of conductivity depending on the configuration and constituents of crystal-like substances. In computational analysis, researchers have critically controlled these variables, unlike empirical analyses in the presence of dopants and flaws, which are susceptible to the manufacturing environments and hard to describe. To address issues relating to the initiation of force fields for processes in which charge transfer and polarizability are applicable, ab initio molecular dynamics (MD) simulations are of interest, and scholars have broadly utilized them. However, due to the somewhat low conductivity of SSE materials (always not up to 10^{-3} S·cm^{-1}), it is frequently computationally expensive, besides being somewhat impractical, to essentially quantify the estimation of the conductivity of the materials considered in ambient condition by means of ab initio MD computations [15,16]. An approach to resolve this difficulty is to compute the coefficient of diffusivity at elevated temperatures and then apply the Arrhenius equation to speculate the ionic diffusivity at ambient temperature [15,17]. Meanwhile, scholars expect the relative errors in the eventual conductivity value to be substantial, since the high-temperature data for statistical errors engender more relative errors for such estimated values [15]. So, they need to introduce proxy computational approaches which give definitive approximates of the diffusion coefficient to review the processes. Researchers have reported that ab initio non-equilibrium molecular dynamics (AI–NEMD) simulations is useful in this premise. Although they used AI–NEMD simulations for the ionic conductivity of LiBH$_4$ in a review [18], they did not juxtapose the outcome of that review with that from ab initio equilibrium molecular dynamics (AI–EMD) simulations. So, it is demonstrated that AI–NEMD simulations permit good estimates for the coefficients of diffusivity acquired for materials with diffusivities not precisely ascertained by AI–EMD computations [19]. They used these values to establish materials such as SSE for recognizing the objective for modifications in experimentations.

An argyrodite structure (for example, Li$_6$PS$_5$Cl) is a cubic crystal structure possessing $\overline{F}43$ m, space group number 216 [20,21]. The literature have examined argyrodite supercell assumed from two-unit cells each containing four Li$_6$PS$_5$Cl, with 104 atoms (eight simple unit cells) for the entire computations of the argyrodite material. The regulation of the unit cell lattice parameters was conducted through energy minimization, so the dimension is 10.08 Å × 10.08 Å × 10.08 Å (Å = Angstrom). From Figure 1, Li-ions fill 48 h Wyckoff positions, where there is a distribution of S atoms on 4a and 4c sites. Again, there is an attachment of S in the 4a areas to the P atoms (4b areas) framing PS_4^{3-} (labeled S1), surrounded by Li-ions is the S as S^{2-} in 4c sites (labeled S2), so the Cl$^-$ also get dispersed in 4a sites. There is a formation of pure and defective Li$_6$PS$_5$Cl structures gleaned from its space lattice. When discussing the creation of the Li$_5$PS$_4$Cl$_2$ structure, it is crucial to replace the S2 S ions with Cl ions and achieve charge balance by removing one of the lithium-ions from the structure surrounding each individual S2 [21].

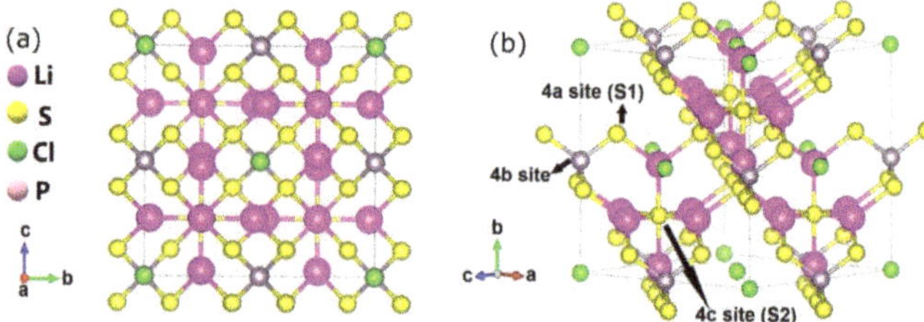

Figure 1. Sample of Li$_6$PS$_5$Cl lattice structure. Purple = lithium, yellow = S, green = Cl, and light purple = P atoms (at the middle of the PS_4^{3-} ion). Adapted from Baktash et al. [19].

In Sections 2–4, we explain the general modeling approaches for argyrodite materials. They also contain comparisons of common argyrodite compounds. Sections 5 and 6 are for some particular argyrodite compounds of high interest amongst researchers due to their electrochemical properties. Ab initio molecular dynamics modeling specifics and machine learning for the argyrodites are presented in Section 7. Sections 8 and 9 are for discussions and conclusions. Therefore, this work reviews the previous work conducted on the modeling of argyrodite electrolytes materials with machine learning (ML) predictive models as speculating the utilization of lithium as a peculiar cathodic substance.

2. Modeling of Argyrodite Materials

The determination of diffusion coefficients of argyrodite materials is essential; however, it is relatively problematic. The diffusion coefficients are calculated using AI–NEMD and AI–EMD simulations for the high diffusive SSE to illustrate an obtained precision. Moreover, the AI–EMD is terribly slow to investigate diffusion coefficients, whereas it is viable and timely to get them using AI–NEMD simulations. Therefore, utilizing AI–NEMD simulations for the coefficient of diffusivity of Li-ions in argyrodite materials (Li_6PS_5Cl and $Li_5PS_4Cl_2$), Baktash et al. obtained two potential electrolytes for ASSB [19]. These computations prove Li-ions diffusion coefficient in $Li_5PS_4Cl_2$ is greater than other promising SSEs, making it bespoke for future technologies. Small disorders and voids were considered as possible explanations for the variability observed in experimental data. Remarkably, inculcating Li vacancies and disarray into the system tends to improve Li_6PS_5Cl ionic conductivity.

Again, amongst the ASSBs, sulfide-based electrolytes, as a result of their being electrochemically stable, noble mechanical characteristics and ionic conductivities are one of the promising candidates, more so than other probable SSEs [22–25]. The Li-argyrodites are built on Li_7PS_6, possessing substantial ion conductivities of 10^{-5}–10^{-3} $S{\cdot}cm^{-1}$ under ambient condition [26]. Considering the lithium-argyrodites model peak temperature point, its ionic conductivity as well, is higher at an elevated temperature point, high conductivity argyrodite models, such as Li_7PS_6, are unstable at ambient temperature. The literature reported that the creation of lithium vacancies and institution of halogens into the Li_7PS_6 model make it possible to produce configurations of Li_6PS_5X (X = Cl, Br, and I) [1] which are constant at ambient temperature with higher conductivity [27,28]. Empirical reports prove Li_6PS_5X (X = Cl, Br, and I) to possess identical lattice configurations of space group, $\overline{F}43$ m, with Li_7PS_6. As a matter of fact, Li_6PS_5Cl and Li_6PS_5Br ionic conductivities at ambient temperature are sufficiently high for battery technology [29,30]. Moreover, halides disorder effects are researched [28] as investigations prove that Li_6PS_5Cl have an ionic conductivity of about 10^{-3} $S{\cdot}cm^{-1}$ at ambient temperature and electrochemical stability up to 7 V versus Li/Li^+ [3,30,31]. Computations have suggested that more halogens and Li vacancies give rise to greater ionic conductivities, and it is speculated that $Li_5PS_4Cl_2$ could be a replacement [21]. Many experimental and computational overviews have been conducted for Li_6PS_5Cl, but because of numerous difficulties, there is not enough full absolute knowledge of the mechanism of the diffusivity of this specimen, and the conjecture for its conductivity varies over many orders of magnitude [16,17,21,27,30–32]. Researchers have reviewed a lot of superionic materials to be potentially lithium solid electrolytes [33,34]. While oxides and phosphates are inelastic and stiff, Li_3PO_4 [3] tend to be flexible, simply formed, and densified, with modulus of rigidity reduced 5 to 10 times [35,36], showing good ionic conductivity of about 25 $mS{\cdot}cm^{-1}$ at ambient temperature [37–46].

Amongst the phases provide in Table 1, the last-mentioned phase is considered the most stable in connection with lithium metal, since a phase consisting of Li_2S [1], Li_3P and LiX (X = Cl, Br) develops at a relatively low order when in connection with the lithium that works as an in situ passivation interphase [47]. Some innovative ASSBs cells utilized lithium-argyrodite as SSEs accompanied by various cathodic and anodic arrangements [30,48–54]. Nonetheless, scientists have designed all the Li-argyrodite electrolytes investigated so far, and their modeling is performed by a conventional mechanical milling machine followed by the application of heat. Ball-milling exudes substantial energy and makes the synthesis hard to measure and thereby not feasible to SSB formation. Solution-engineered processes did not alone solve these issues, but improbably it also decreases the ensuing temperature of the heat treatment coupled with time. Solution-based methods were engineered for Li_4PS_4I [55], β-Li_3PS_4 [1,56], $Li_7P_2S_8I$ [57], and $Li_7P_3S_{11}$ [58] structures, alongside differing levels such as the purity of the phase and ionic conductivity of the result. The literature delineated Li_6PS_5Cl argyrodite, requiring a dissolution−reprecipitation procedure arising out of a solution of ethanol [32]. While mechanical milling measure is essential, as the ionic conductivity of the specimen reported is 1.4×10^{-2} $mS{\cdot}cm^{-1}$, which is two orders of magnitude less than the traditional solid-state formation technique, an exact and valid "all-solution" argyrodite Li_6PS_5X formation stands out as a question.

Table 1. Lithium superionic conductor phase [1,8].

Phase	Ionic Conductivity (IC) (mScm^{-1})	Activation Energy (E$_a$) (eV)	Reference
$Li_{3.25}Ge_{0.25}P_{0.75}S_4$	2.2	0.21	[1,37]
$Li_{10}GeP_2S_{12}$	12	0.25	[1,3,38] and derivatives [39,40]
$Li_{9.54}Si_{1.74}P_{1.44}S_{11.7}$	25	0.24	[41]
$Li_7P_3S_{11}$	17	0.18	[1,42–44]
Li_6PS_5X (X = Cl, Br)	~1	0.3−0.4	[31,45,46]

Reprinted (adapted) with permission from *ACS Energy Lett.* **2019**, *4*, 1, 265–270 [8]. Copyright 2018 American Chemical Society.

It was shown that a direct solution-engineered procedure of Li_6PS_5X (X = Cl, Br, I) argyrodite [1] as well as $Li_{6-y}PS_{5-y}Cl_{1+y}$ SSEs possess good lithium-ionic conductivities, which is about 2.4 mS·cm^{-1}, with reference to ambient temperature against chlorine and bromine phases. Ionic conductivity is more for an integrated Cl and Br, and chlorine-full phases, yielding about 4 mS·cm^{-1}.

3. Single Halides Argyrodites

The articulation in Kraft et al. [28] proves that solid-state engineered Li_6PS_5Cl and Li_6PS_5Br show an exact bulk of LiX and Li_3PO_4 dopants. Yubuchi et al. [59] reported the results culminating from the composite synthesis showing a more dopant level, attributed to the reactivity of ethanol and PS_4^{3-} units, or slight moisture in the ethanol. Zhou et al. [8] used a related solution synthesis approach to produce a pure Li_6PS_5I phase with an imprint of lithium thiophosphate (Li_3PO_4) impurity, albeit with Li_6PS_5I argyrodite low ionic conductivity, which agrees with reviews from Pecher et al. [60]. However, the iodide crystallization explains a wide relevance of this solution synthesis procedure to the lithium-argyrodite species. Researchers represented the SEM images of grounded Li_6PS_5X (X = Cl, Br) materials to interpret the densified characteristics of the crystallite masses, and when processed into ASSB, they become incredibly useful [8]. Highly ductile sulfide solid electrolytes give good contact at the grain boundaries than oxides (which are always essentially brittle), even pressed parts without consideration of sintering result in multiple solid electrolyte phases [43]. The reported solution engineered ionic conductivities of Li_6PS_5X SSEs were determined by electrochemical impedance spectroscopy (EIS) in (stainless steel = SS) SS/Li_6PS_5X/SS arrangement at ambient temperature. Researchers reported the resistivities, 46 and 34 Ohm, for Li_6PS_5Br and Li_6PS_5Cl at ambient temperature, making the total conductivities 1.9 mS·cm^{-1} and 2.4 mS·cm^{-1}, discretely. These values correspond with that of solid-state engineered Li_6PS_5Br and Li_6PS_5Cl argyrodites of about 1 mS·cm^{-1} and 3 mS·cm^{-1} apiece [28,31,46]. This shows that the dopant effects in the processed materials do not lower the Li-ions conductivity in the exceptionally fine electrolytes. It is noteworthy that these elements are especially improbable to enhance the gross ionic conductivity as the densified solid electrolyte pressed pieces enable good interaction amid Li_6PS_5X lattices and allow ample lithium ion straining pathways for transport [28]. Iodide argyrodite conductivity was reported to be 2×10^{-3} mS·cm^{-1}, which tallies with the low values often published for this phase (i.e., 4×10^{-4} mS·cm^{-1}) [60]. There is the measurement using a direct current (DC) polarization for the electronic conductivities of the two materials (Li_6PS_5Cl and Li_6PS_5Br) of the SS/Li_6PS_5X/SS symmetric cells at ambient temperature. Zhou et al. [8] estimated the direct current electronic conductivities as 5.1×10^{-6} mS·cm^{-1} for Li_6PS_5Cl and 4.4×10^{-6} mS·cm^{-1} for Li_6PS_5Br. Both values are six orders of magnitude smaller than their individual ionic conductivities. Proving that

Li_6PS_5Br and Li_6PS_5Cl argyrodite electrolytes obtained from synthesis are productively unalloyed ionic conductors.

4. Mixed Halide Argyrodites

Formerly researched mixed halide argyrodites Li_6PS_5X (X = $Cl_{0.75}Br_{0.25}$, $Cl_{0.5}Br_{0.5}$, and $Cl_{0.25}Br_{0.75}$) present substantial ionic conductivities when contrasted with single-halide phases [28], enabling the study of the molecular modeling of argyrodites. Zhou et al. [8] further showed the X-ray diffraction (XRD) patterns after heat treatment for the product, as the identical values reported for the single halide composition, especially the pure argyrodite phase, is available as an important crystalline phase with the presence of dopants of Li_3PO_4, Li_2S, and LiX (X = Cl, Br). Additionally, the lattice parameters from Rietveld refinements increase directly from x = 0–1 in $Li_6PS_5Cl_{1-x}Br_x$. For solid solutions, as explained by Vegard's principle, the disordered Cl/Br ions were constructed in the model, which demonstrates the argyrodite phases. They used EIS to measure the ionic conductivities and recorded an ionic conductivity of 3.9 mS·cm^{-1} for $Li_6PS_5Cl_{0.5}Br_{0.5}$, and using a DC polarization, they reported particularly low electronic conductivities. The ionic and electronic conductivities are summarized in Table 2.

Table 2. Compendium of the ionic and electronic conductivities of lithium-argyrodites primed by synthesis method [1,8].

S/N	Material	IC (mS·cm^{-1})	EC (mS·cm^{-1})
1.	$Li_{5.5}PS_{4.5}Cl_{1.5}$	3.9	1.4×10^{-5}
2.	$Li_{5.75}PS_{4.75}Cl_{1.25}$	3.0	2.6×10^{-5}
3.	Li_6PS_5Br	1.9	4.4×10^{-6}
4.	$Li_6PS_5Cl_{0.25}Br_{0.75}$	3.4	1.1×10^{-5}
5.	$Li_6PS_5Cl_{0.5}Br_{0.5}$	3.9	1.4×10^{-5}
6.	$Li_6PS_5Cl_{0.75}Br_{0.25}$	3.2	3.7×10^{-6}
7.	Li_6PS_5Cl	2.4	5.1×10^{-6}

Reprinted (adapted) with permission from *ACS Energy Lett.* **2019**, *4*, 1, 265–270 [8]. Copyright 2018 American Chemical Society.

Previously reported density functional theory (DFT) MD simulations further increased the Li ion conductivity through the sulfur/halide ratio adjustment [21], shown by the XRD scales of $Li_{5.75}PS_{4.75}Cl_{1.25}$ and $Li_{5.5}PS_{4.5}Cl_{1.5}$ formulated by the solution-engineered synthesis approach. Lithium-argyrodite electrolytes are available as the prime products, with dopants of Li_3PO_4 contrasted with Li_6PS_5X, little Li_2S and carefully selected LiCl. EIS reported surging ionic conductivities with a greater Cl-to-S ratio for $Li_{5.75}PS_{4.75}Cl_{1.25}$ and $Li_{5.5}PS_{4.5}Cl_{1.5}$, respectively (see Table 2). The incremental ionic conductivities and XRD scale show a replacement of S with Cl, introducing Li-ions voids for the argyrodite structure, thus enhancing the ionic conductivity. However, because of the homogeneous array and X-ray smattering conditions of S and Cl, advanced neutron divergence work are important to differentiate the order of S/Cl disarray, Li vacancies and positioning in the argyrodite processes.

Again, Zhou et al. [8] considered $TiS_2/Li_{11}Sn_6$ ASSBs prototype using Li_6PS_5Cl as an SSE formulated between 1.5V and 3.0V vs. Li/Li$^+$ at 30 °C. Obviously, these cells utilized lithium alloy as a negative electrode due to their toughness with the goal of measuring the qualities of argyrodite. Prospects will make use of lithium as a discrete cathodic substance. Figure 2 presents two discharge–charge voltage depiction for Li_6PS_5Cl cells formed through the traditional solid-state approach (Figure 2a) and the solution-synthesized process (Figure 2b).

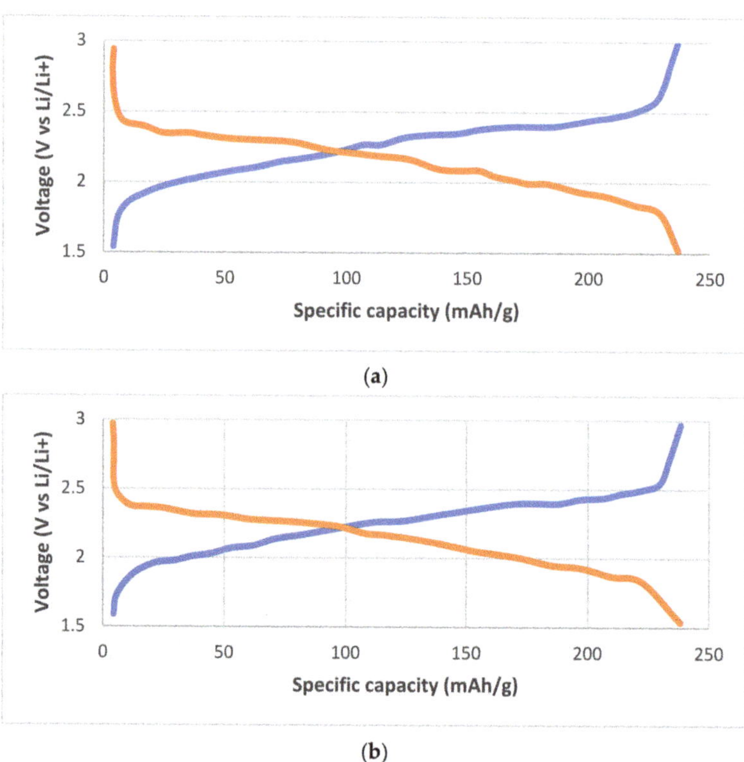

Figure 2. Comparison of Li_6PS_5Cl ASSB by (**a**) solid-state procedure (**b**) solution-synthesized procedure of $TiS_2/Li_6PS_5Cl/Li_{11}Sn_6$ cells at current density, 100 $\mu A \cdot cm^{-2}$ (equivalent to 0.11 charge rate). Reprinted with permission from *ACS Energy Lett.* **2019**, *4*, 1, 265–270 [8].

The Li_6PS_5Cl cell produced by the solution-synthetic method presents an ideal capacity of 239 mAh·g^{-1} and presents no change from cells making use of derived SSE, suggesting that the slight scum produced within the solution-synthesized pathway are negligible in comparison with the overall performance of the cell.

5. Ionic Conductivity of $Li_5PS_4Cl_2$

Utilizing AI-EMD and AI-NEMD simulations at 300 K, 600 K, 800 K, and 1000 K, the diffusion coefficient for $Li_5PS_4Cl_2$ could be calculated. So, this process advances a specific collation with the accuracy of AI-EMD and AI-NEMD techniques and with extrapolated calculation at 300 K using the Arrhenius equation. There is an assessment of the AI–EMD simulations at 300 K, 600 K, 800 K and 1000 K for 45 ps, 40 ps, 20 ps and 20 ps, discretely. Researchers have calculated the MSD [16] of the Li-ions, values for ten models, and statistical errors at various temperatures using the results of these AI-EMD simulations. They examined the trajectories in each case to ensure that the simulation times (450 ps–200 ps) were enough to distinguish the Li-ions diffusion further from their starting points. Figure 3a–c show the MSD of the lithium-ions with time dependence at 300 K, 600 K and 800 K for the ten separate trajectories. In a diffusive motion, the MSD heightens with time, and the gradient is associated with diffusion co-efficient from Equation (1) [61].

$$D_s = \lim_{t \to \infty} \frac{1}{6Nt} \sum_{i=1}^{N} \langle (\Delta r_i(t))^2 \rangle \tag{1}$$

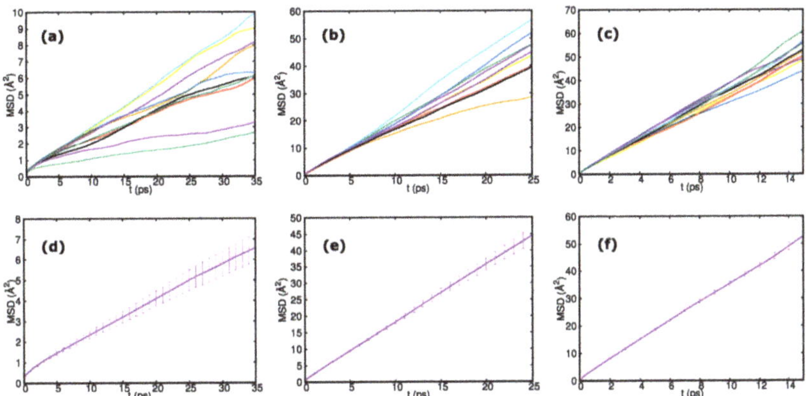

Figure 3. Lithium-ions mean square displacement (MSD) with time. Results for ten iterative computations of $Li_5PS_4Cl_2$ at (**a**) 300 K, (**b**) 600 K, (**c**) 800 K, and MSD of the ions with respect to time with the error bars for ten iterative simulations considering (**d**) 300 K, (**e**) 600 K, (**f**) 800 K. Adapted from Baktash et al. [19].

(See nomenclature for terms definition.)

Figure 3d–f show the derivation of the MSD of 10 models involving the error bars where the coefficients of self-diffusion and the proportion of the conductivity of the material with an approximate computational accuracy. While noticing diffusive behavior, the value of the MSD at 35 ps at 300 K is only about 6 ^{2}. The conductivity and coefficient of self-diffusion values are shown in Table 3, respectively.

Table 3. Computation using AI–EMD and AI–NEMD simulations for self-diffusion coefficient with concurrent conductivity of $Li_5PS_4Cl_2$ at specific temperatures.

Temp. (K)	AI–EMD		AI–NEMD	
	D_s (cm^2 s^{-1})	σ (S·cm^{-1})	D_s (cm^2 s^{-1})	σ (S·cm^{-1})
300	2.9×10^{-6}	0.35	3.3×10^{-6}	0.4
600	2.9×10^{-5}	1.80	2.9×10^{-5}	1.8
800	5.6×10^{-5}	2.50	5.2×10^{-5}	2.4
1000	8.9×10^{-5}	3.20	8.9×10^{-5}	2.9

Adapted from Baktash et al. [19].

For SSEs with minimal conductivity, such as those of about 10^{-3} S·cm^{-1} or less at ambient temperature, it is complex to compute an accurate conductivity at ambient temperature with ab initio simulations. In order to find a solution, it was specified that the conductivity of the material at peak temperatures may be calculated using a straightforward method, and that the values for the ambient temperature can be extrapolated to produce an approximation of the conductivity value. [18,61,62]. An Arrhenius representation for the conductivity of Li-ion data computed from AI-EMD simulations in $Li_5PS_4Cl_2$ from MSD for various temperatures is illustrated by Figure 4.

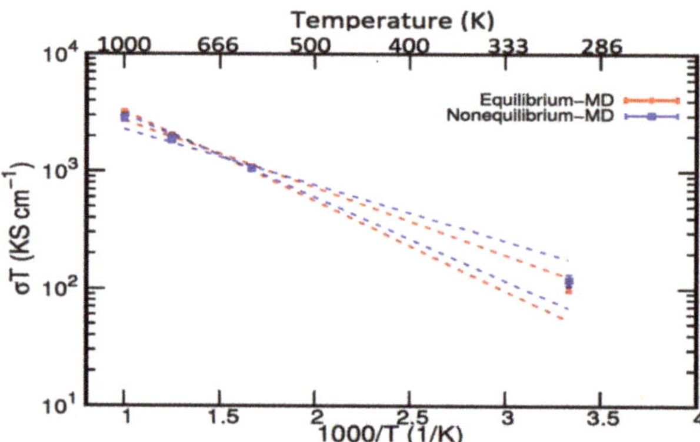

Figure 4. Arrhenius plot for Li$_5$PS$_4$Cl$_2$ conductivity. Outcomes of AI–EMD simulations and AI–NEMD simulations at specific temperature. The error ranges for extrapolation estimated data at 600 K and 800 K to 300 K are displayed by AI-EMD and AI-NEMD, respectively. Adapted from Baktash et al. [19].

From MSD [17] computations, the ionic conductivity of Li$_5$PS$_4$Cl$_2$ at 300 K is 0.35 ± 0.05 S·cm^{-1}. Figure 4 presents a close look into the values at 600 K and 800 K, the extrapolated conductivity presents an ionic conductivity ranging from 0.17–0.37 S·cm^{-1} at 300 K. This works well with the computed results. However, the statistical error for the elevated temperature values is minute when related to the results. Moreover, correlative error at decreased temperature is sizeable because of their extrapolated values. Additionally, the extrapolated result at 1000 K gave values from 0.17–0.33 S·cm^{-1} at 300 K, thereby realizing a minimal swap through an extra-elevated temperature since it requires a substantial temperature difference to advance a discrete value in the temperature reciprocal.

Aeberhard et al. [18] employed an alternate method based on AI-NEMD simulations to calculate the self-diffusion coefficient of hexagonal LiBH4 at 535 K, which is different from the method used in ab initio simulations to find the IC of SSE. To prove the accuracy of Baktash et al. [19], AI–NEMD simulations approach is employed to speculate the Li$_5$PS$_4$Cl$_2$ coefficients of self-diffusion and conductivities at 300 K, 600 K, 800 K and 1000 K, so these values are juxtaposed with those from AI–EMD MSD calculations.

Calculating conductivity at various fields is necessary to identify the linear response domain for ionic conductivities calculations from AI-NEMD simulations. The EMD and NEMD procedures reported results work in the range of one standard error. There is an evaluation of the conductivity of the extrapolated specimen at 300 K from elevated temperatures. Distinguishing the lines in Figure 4 using comparable complete simulation times at 300 K, 600 K and 800 K for the NEMD and EMD computations, the NEMD simulations statistical errors values equate that of EMD approach. This is true for both direct consideration and extrapolated elevated temperatures values at 300 K.

Li$_5$PS$_4$Cl$_2$ Collective Diffusion Coefficient

Scholars used the Nernst–Einstein models to connect the conductivity of a material to its coefficient of self-diffusion. However, if the sample of interest diffusing atoms tends to be stationary at the time of diffusion, on top of that, it is an ineffective estimation, and therefore, it is advisable to use the coefficient of collective diffusion [63]. When the diffusing atoms transpose individually, the coefficients of self and collective diffusion coincide, but they diverge when they do so collectively.

$$D_c = \lim_{t \to \infty} \frac{1}{6Nt} \langle \left(\sum_{i=1}^{N} r_i(t) - \sum_{i=1}^{N} r_i(0) \right)^2 \rangle \tag{2}$$

(See nomenclature for terms definition.)

In order to determine if this calculation is adequate for the systems under consideration, Evans et al. [61] computed the $Li_5PS_4Cl_2$ coefficient of collective diffusion using Equation (2) and compared it to the coefficient of self-diffusion. They picked the most diffusive pure system because it is challenging to make deductions when the statistical error is considerable, and since this statistical error for its collective diffusion computation is substantially higher than that of self-diffusion estimations. A report from Baktash et al. [19] compares the MSD discrete sample simulations for $Li_5PS_4Cl_2$ at 800 K for the ions and their center of mass, but the error bar for the ion MSD remained low. This holds true since each ion in the sample can offer an autonomous contribution towards MSD. Therefore, the MSD (self and collective diffusion coefficients) computed in both ways agree. The agreement aligns with individual performance of the entire lithium-ions in $Li_5PS_4Cl_2$ during the diffusion process, making it easy to compute the conductivity by the coefficient of self-diffusion.

6. Li_6PS_5Cl Ionic Conductivity

Keil et al. [64] reported that for diffusive motions, simulation times vary inversely with the coefficient of diffusion. For low conductivity models, such as EMD simulations, it is hard to obtain precise values, and at some points, there is no activity of ions in a practicable timescale between the areas in the electrolyte.

An analysis of $Li_5PS_4Cl_2$ has shown that valid findings were achieved by coalescing NEMD computations under elevated temperatures with assumed values at minimal temperatures. So, a method of sustaining the range of materials to compute the conductivity by providing an approach of when the conductivity is very minimal to allow the use of EMD simulations has been reported. Baktash et al. [19] proved this tender by computing Li_6PS_5Cl ionic conductivity by NEMD simulations. Scientists have utilized both AI–EMD and AI–NEMD simulations with the objective of computing the conductivity and recognizing the technique of diffusion in pure Li_6PS_5Cl, Li_6PS_5Cl with sulfur–chlorine disarray (sulfur and chlorine switching locations), and Li_6PS_5Cl with mutually lithium-ion vacancies with sulfur–chlorine disarray.

6.1. Li_6PS_5Cl Pure Diffusion

Six Li-ions move in an octahedral region around an S2 S atom in one of the two different motion types, while a second Li-ion transposes in the center of these octahedral borders in the other. If the skips time is more than the time of simulation, then no conclusion is viable on the diffusion coefficient from the simulation. For AI–EMD simulations of Li_6PS_5Cl, they did not notice jumps between cages in ten independent trajectories of 100 ps at 300 K and 450 K and extraordinarily little, or no jumps were visible for the 600 K trajectories. This proposes a small coefficient of diffusion of the material at 300 K (not up to 10^{-4} S·cm^{-1}) and exemplifies the strain of assessing the result of the conductivity by EMD computations [64]. However, it is a simple exercise to obtain conductivity results utilizing NEMD computations at 600 K and 800 K, where jumps observation is available, and to gauge the conductivity at 300 K using an extrapolation of these results. Baktash et al. [19] envisaged the pure Li_6PS_5Cl conductivity ranging from 10^{-5}–10^{-4} S·cm^{-1}. The period of simulation needed to obtain a corresponding accuracy for the Li_6PS_5Cl conductivity through the EMD method at 600 K and 800 K was restraining as mean and standard error jumps did not take place. Meanwhile, considering another option at elevated temperatures, the extrapolated errors will be much more if the lowest temperature tends to be greater, eventually leading to a disordering of the complex followed by a diffusion coefficient with a corresponding non-Arrhenius behavior. Thus, the NEMD method is a more suitable option.

The pure Li_6PS_5Cl lithium-ion conductivity was formerly established empirically [20,32,65] and computationally [17,21] at 300 K. It is interesting to note that the calculations from earlier publications for pure Li_6PS_5Cl at 300 K differ by five orders of magnitude and are between one and four orders of magnitude distinct from the actual data [19]. They obtained a lower extrapolated data for simulations at 600 K and above. Consequently, reports show no existence of consistent computational approximates of the diffusion coefficient for this system. Under this condition, neither jump frequency of lithium-ions in Li_6PS_5Cl in ambient condition at about 109 s^{-1} [7] demands extrapolation of the elevated temperature and use of AI–NEMD to predict the conductivity at lower temperatures.

6.2. Li₆PS₅Cl: Disorder Effect

The Li_6PS_5Cl high conductivity spotted in several research studies is as a result of the disarray of Cl and S atoms [7,20,66]. To apprehend this, the consideration of a model with disarray based on the empirically scrutinized structure is important. There is a swapping of 25 percent of the Cl ions holding Li_6PS_5Cl 4a areas/sites with S2 ions which occupied 4c areas/sites. EMD simulations values were taken to determine the conductivity of the disarrayed configuration at 600 K and 800 K. Assuming Arrhenius behavior, the empirical values for the conductivity of the disarrayed configuration are relayed in Baktash et al. [19] as squares and triangles [7,31,67,68] with reported computational results [21]. Studying the error bars, the computed result for the conductivity of the dissembled structure, aligns with the contemporary empirical results reported by Yu et al. [7]. As a result, the disorganized structure has a conductivity that is two orders of magnitude greater than the pure Li_6PS_5Cl structure. Accordingly, minor structure changes include replacing four out of the 104 atoms, gives a sizeable altering of the conductivity. This demonstrates that minute voids in the pure Li_6PS_5Cl crystal might result in a range of empirical conductivities. There is the regulation of the diffusion mechanisms of the ordered and disarrayed Li-ions structures at 600 K using AI–EMD simulations (Figure 5). Comparing the methods, it is evident that switching the positioning of Cl and S ions cause a disarray that alters the freedom of the Li-ions. Figure 5a,b show the movement of pure Li-ions in the octahedral regions produced by the PS_4^{3-} (Figure 1), and the paths through the regions show that the energy barrier for diffusion out of the regions is high when compared to the available thermal energy, which is why the pure crystal ionic conductivity is hardly discernible from these AI-EMD simulations.

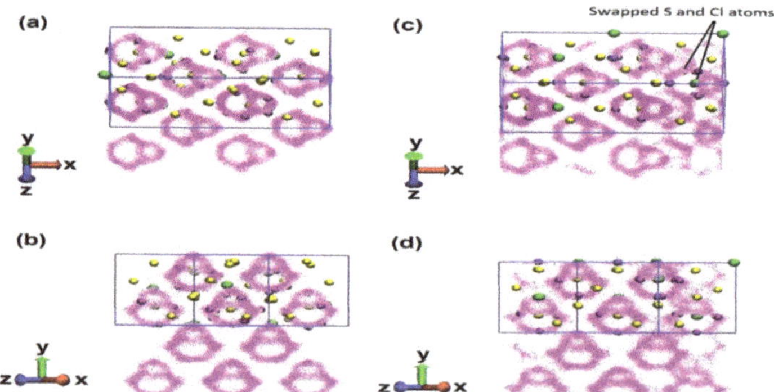

Figure 5. Li ion trajectories views, pure Li_6PS_5Cl (**a,b**) and (**c,d**) for S-Cl disarrayed Li_6PS_5Cl at 600 K observed up to 50 ps. Li (violet), S (yellow), Cl (green) and P (light purple). The diffusion pathways of Li-ions = small violet dots. The apertures illustrate simulation supercell. Adapted from Baktash et al. [19].

While in the disordered structure, an aperture linking the cages is visible, Figure 5c,d. This demonstrated that in the new passage direction, diffusion of lithium-ion barrier energy is miniature compared with other regions and with the size it used to be in the pure structure, as lithium-ions advance into the dissembled area more readily.

6.3. Li_6PS_5Cl: Li-Ions Vacancies and S–Cl Disorder

Yu et al. [66] proposed the empirically formed specimen to possess an empirical formula of $Li_{5.6}PS_{4.8}Cl_{1.2}$ after heat treatment. They believed that the charges were not balanced in the empirical formula presented; however, a formula that comes near to charge balance is $Li_{5.8}PS_{4.8}Cl_{1.2}$. Considering the influence of the fusion of a lithium vacancy, and a few more chlorine ions and fewer S2 ions [67] on the Li_6PS_5Cl ionic conductivity, two Li-ions were removed from the Li_6PS_5Cl supercell (8–unit cells) and two S2 ions were changed with two chlorine ions [45], delivering a novel $Li_{5.75}PS_{4.75}Cl_{1.25}$ structure, more like the formula reported in previous work ($Li_{5.8}PS_{4.8}Cl_{1.2}$). The calculated conductivity of the ions in this structure at 300 K was 0.09 ± 0.03 S·cm^{-1}, which is somewhat higher than the calculated result for Li_6PS_5Cl with disorder of the sulfur and chlorine atoms alone (6×10^{-3}–10^{-1} S·cm^{-1}) and higher than the evaluated value for pure Li_6PS_5Cl solid electrolyte. The approximated result for the ionic conductivity is also greater than that related as 1.1×10^{-3} S·cm^{-1} in Yu et al. [66]. However, sensitivity to voids and disarray means they are not anticipated to match, as a result of the variance in empirical formulas. It is obvious that the diffusion of the structure is improved by lithium vacancies and the substitution of a sulfur ion for a chlorine ion. They were prepared for this increase since the lithium vacancies and the small masses of the chlorine and sulfur ions tend to cause the mobility of the lithium-ions in space, and the disorder caused by the substitution of chlorine ions for sulfur ions may facilitate easy movement.

Table 4 shows a compendium of conductivities of the various type Li_6PS_5Cl structures (pure, disarrayed, and structures with both disarray and lithium vacancies) reported from computational works. From the outcomes of the computations and previous reviews [17,20,21,27,30,32,66], pure Li_6PS_5Cl has a quite low conductivity of 10^{-5}–10^{-4} S·cm^{-1}. However, impurities such as lithium vacancies, ion disorder, and grain boundaries instituted during the synthesis of argyrodite electrolyte would affect their final conductivity.

Table 4. The conductivity of Li_6PS_5Cl and defective materials at 300 K investigated in tens of research and computational studies.

Material	IC (S·cm^{-1})	Reference
Li_6PS_5Cl	1.4×10^{-5}	Experiment [32]
	3.3×10^{-5}	Experiment [20]
	6×10^{-5}–3×10^{-4} *a	AI–NEMD [19]
	6×10^{-5}	Experiment [65]
	0.29	Computation (MSD) [21]
	0.05 (0.16)	Computation (Jump) [21]
	2×10^{-6}	Computation (MSD) [17]
$Li_{5.6}PS_{4.8}Cl_{1.2}$	1.1×10^{-3}	Experiment [66]
$Li_{5.75}PS_{4.75}Cl_{1.25}$	6×10^{-2}–1.2×10^{-1} *a	[19]
Li_6PS_5Cl with chlorine and sulfur disorder	1.9×10^{-3}	Experiment [67]
	4.96×10^{-3}	Experiment [7]
	3.38×10^{-3}	Experiment [68]
	0.26	Computation (MSD) [21]
	0.89 (1.29)	Computation (Jump) 21]
	6×10^{-3}–0.1 *a	AI–EMD [19]

*a—extrapolation ranges at 800 K and 600 K. Adapted from Baktash et al. [19].

7. Methods of Argyrodite Electrolyte Modeling

Previous works have reported that MD simulations were used to examine the characteristics of ionic liquid (IL) electrolytes close to cathodic electrodes [69]. It is pertinent that we describe that the interfacial structure of IL electrolytes was first considered with MD simulation, as reported by Lynden-Bell [70] and Lynden-Bell et al. [71]. Some novel advancement on the theory of electric double layer, and computer simulation of interfacial structures of IL., new coarse-grained ILs and foregoing discoveries on MD are outlined by

Fedorov et al. [72]. With respect to other work conducted, we decided to streamline our reviews on the modeling of ab initio MD simulation in the ensuing section.

7.1. Ab Initio MD Simulations

Ab initio Born–Oppenheimer MD simulation was carried out by the CP2K (Quickstep package) [73,74] with an improved version of the same that utilized the AI-NEMD procedure. Other research work in the literature inculcated the Vienna Ab initio Simulation Package (VASP) [3] for this same simulation [1,17,75]. Researchers preferred the PBE generalized gradient approximation (GGA) for the density functional theory [3,14] exchange-correlation functional. Van der Waals interactions reported the corrections of the DFT-D3 [76], an approach for many works in the literature. The pseudopotentials of Goedecker, Teter and Hutter (GTH–POTENTIAL) were used [77]. They usually select the DZVP–MOLOPT–SR–GTH [78] basis set as an optimization tool for molecular gas properties and condensed-phase computations. Researchers used the Gaussian and plane-waves (GPW) basis [79] and a cutoff energy (in Rydberg) to further the task. Researchers used a $1 \times 1 \times 1$ k-point mesh (Γ point) in all computations, and initial optimization computations showed the use of grid-initiated errors not up to 0.01 Å for lattice parameters and errors not up to 0.06 eV for the energy per unit cell. The optimized lattice constants, 10.08 Å, used are 2.5% more than the empirically described results for Li_6PS_5Cl and compared with previous computational reports [17,21].

7.2. Equilibrium and Nonequilibrium Models of Motion

EMD and NEMD simulations assess the coefficients of the diffusion of materials in the NVT ensemble (constant-temperature, constant-volume). The models of motion for the AI–EMD simulations are [61,80]:

$$q_i = \frac{p_i}{m_i} \tag{3}$$

$$p_i = F_i - \alpha p_i \tag{4}$$

where q_i = position, p_i = momentum, m_i = mass of atom i, F_i = inter-atomic force on atom i, α = Nosé–Hoover thermostat couples of the atoms. A chain thermostat with the chain number of 3 (typical for CP2K) for EMD computations was employed [19,61,81]:

$$\alpha = \frac{1}{Q_1}\left(\sum_{i=1}^{N} \frac{p_i^2}{m_i} - gk_BT\right) - \alpha\alpha_2 \tag{5}$$

$$\alpha_2 = \frac{Q_1\alpha^2 - k_BT}{Q_2} - \alpha_2\alpha_3 \tag{6}$$

$$\alpha_3 = \frac{Q_2\alpha_2^2 - k_BT}{Q_3} \tag{7}$$

In the above equations, Q_1, Q_2, and Q_3 = coefficients of friction, all given the same inputs of three in the equilibrium simulations, k_B = Boltzmann's constant, T = target temperature and g = degrees of the system freedom. The non-equilibrium approach applied to resolve the coefficient of diffusion for lithium ion is the color-diffusion algorithm that is broadly functional in classical MD simulations [80–82]. The equations of motion for ions utilizing this line of instruction are given in Equations (3) and (4); however, the equation of motion for the momentum of the lithium-ions is given. Equation (4) was modified to incorporate a force brought on by a color field, F_c. [80]:

$$p_i = F_i + c_iF_c - \alpha p_i \tag{8}$$

where c_i = color charge of lithium ions. The summation of the color charge = 0, ensuring that no drift of moment occurs in the system, but the selection of these charges is random. There is always +1 color label on half of the lithium-ions and −1 color charge on the other half. PS_4^{3-}, S^{2-} and Cl^- are not directly affected by the color field, and the color charges do not affect how the Li^+ interact with one another; rather, the color field simply affects the field response. Because a chain thermostat is inappropriate for non-equilibrium simulations, it was not applicable for NEMD computing [83].

Using the velocity Verlet technique in the 1 fs time step, it is crucial to extend the EMD and NEMD equations of motion. Using the Nernst–Einstein equation and the coefficients of self-diffusion, researchers may determine the ionic conductivity of any specimen, σ [17,21,80]:

$$\sigma = \frac{ne^2 Z^2}{k_B T} D_s \tag{9}$$

where n is the ion density of Li, e is the elementary electron charge, and Z the valence of Li.

Using the NEMD simulations, the self-diffusion coefficient is derived from the computation of the color current by the color field. Remarkably, this approach determines the self-diffusion coefficient of an SSE with minimal conductivity. The color current is [18]:

$$J_c(t) = \sum_{i=1}^{N} c_i v_i(t) \tag{10}$$

where v_i = velocity of ith lithium ion. At low fields, the color current in the direction of the field ($J_c = J_c \cdot F_c/|F_c|$) varies directly to the field, $F_c = |F_c|$, when the systems are in a steady state, and then for the color charges reported in Evans et al. [61]:

$$D_s = \frac{k_B T}{N} \lim_{t \to \infty} \lim_{F_c \to 0} \frac{J_c}{F_c} \tag{11}$$

where N = number of Li-ions subject to a color field. Using a disparate choice of color charges, we substitute N in Equation (11) with $\sum_i^N c_i^2$. The system and condition, including temperature, will have an impact on the field's outcome if there is a linear change between the color current and applied field. Baktash et al. [19] conducted simulation in practice to calculate the critical field. So, to obtain values with the least statistical error, maximum field for which linear response occurs is better utilized. In ergodic order, researchers substituted the color current ensemble average in Equation (11) with a time average, using [19]:

$$\langle J_c \rangle = \lim_{t \to 0} \frac{1}{t} \int_{t_0}^{t_0+t} J_c(s) ds = \lim_{t \to \infty} \frac{1}{t} \sum_{i=1}^{N} c_i \, \Delta r_i(t) \tag{12}$$

In the horizontal direction, the self-diffusion coefficient is [77,79]:

$$D_s = \frac{k_B T}{N} \lim_{t \to \infty} \lim_{F_C \to 0} \frac{\sum_{i=1}^{N} c_i \Delta x_i(t)}{t F_C} \tag{13}$$

The field accumulates force to the molecules in the orientation of the field assuming a corresponding influence to diminishing the activation energy bar for diffusion [84].

The predicted time for a specific leap, if the diffusion method is synthesized as a jump process, will increase with the size of the activation bar. As a result, using the field discharge permits systems with lower diffusion coefficients for a given simulation duration. This proves that there is a great edge for NEMD computations for systems with minimal coefficients of diffusion. Since the ionic conductivity of SSEs depends on temperature, researchers used ionic conductivities at elevated temperatures calculated from MD simulations to specify the coefficients of diffusion of the electrolytes at reduced temperatures, considering the Arrhenius equation [1,17]:

$$D = D_0 e^{-E_a/(k_B T)} \tag{14}$$

where E_a = activation energy, and D_0 = diffusion pre-exponential factor. Using Equation (9), we express this in terms of the ionic conductivity of the process [19,85],

$$\sigma T = \frac{ne^2 Z^2}{k_B} D_0 e^{-E_a/(k_B T)} \tag{15}$$

Note that for systems where only the Li-ions are moving and transpose separately, the conductivity computed from the Einstein equation for the self-diffusion, Equation (1), which is written as a Green–Kubo equation, [80–82].

$$\sigma = \frac{ne^2 Z^2}{k_B T} D_s = \frac{e^2 Z^2}{3 V k_B T} \int_0^\infty \langle \sum_{i=1}^N v_i(t) \cdot v_i(0) \rangle dt \tag{16}$$

Furthermore, if the theory of independency of motion of the Li-ions is open, then computed from the coefficient of collective diffusion and the Green–Kubo relationship as [81,82],

$$\sigma = \frac{ne^2 Z^2}{k_B T} D_c = \frac{e^2 Z^2}{3 V k_B T} \int_0^\infty \langle (\sum_{i=1}^N v_i(t) \cdot (\sum_{j=1}^N v_j(0)) \rangle dt \tag{17}$$

The Einstein expression used in Zhou et al. [8] avoids problems correlated with convergence of the time-correlation functions in Equations (16) and (17). Importantly, ab initio MD simulations of conductivity has been reported in the literature, and it has been emphasized that that Green–Kubo equation provides regular results using integer charges and velocities of distinct sites, as shown in Baktash et al. [19], and the mean-square dipole displacement computed using Born charges of each atom [86].

7.3. Machine Learning (ML) Models

The machine learning models provide an approach to juxtapose the atoms in a large molecule and their neighboring counterpart, giving a representation of their bonding and interaction. Again, ML makes it easy to make projections for the energies of large molecules or the disparities between the low-accuracy and high-accuracy computations. It is pertinent to add that ML is appropriate to the search for stable and harmless electrolytes for LIBs [87]. Additionally, ML provides solutions to mitigate orders of magnitude in calculations and big data analysis [88]. A lot of models and equations on ML were extensively reported in Refs. [88,89] to calculate DFT using VASP for the total energies and electron densities.

Partial density states that computations for $LiPS_4$ and $LiPO_4$ can be computed with the form [85]:

$$N^a(E) \equiv \frac{1}{\sqrt{\pi}\Delta} \sum_{nk} f_{nk}^a W_k e^{-(E-E_{nk})^2/\Delta^2} \tag{18}$$

The smearing factor (Δ) is always chosen to be 0.1 eV. W_k is the Brillioun zone sampling factor. f_{nk}^a is the weighing factor, n is the band index, and k is the wave vector.

8. Discussion

We reviewed the investigation of the diffusion technique of lithium-ions in Li_6PS_5Cl and $Li_5PS_4Cl_2$ and their coefficients of diffusion. They used the assessment of the EMD and NEMD methods to estimate the ionic diffusion of these materials, whilst their ionic conductivity is about 10^{-3}–10^{-2} S·cm^{-1}. At any time when the ionic conductivity of these samples is low, the EMD simulation time required for short and simple results is currently infeasible. However, it was shown that it is viable to use NEMD to calculate coefficients of diffusion of SSEs with conductivities in the range 10^{-6}–10^{-4} S·cm^{-1}. The consensus on the results of these techniques has shown that if conductivity is increased sufficiently, both can provide reliable measurements of conductivity. However, for low-conductivity resistance SSEs that are close to ambient temperature, NEMD simulation should be used.

As the diffusivity decreases, the advantage of NEMD calculations over EMD calculations grows. This is due to the fact that the diffusion coefficient has an inverse relationship with the amount of time needed to explore a material in equilibrium simulations [65], but the applied force field also affects the NEMD simulation. It is important to perform different NEMD simulations on fields in distinctive directions to calculate the diffusivity, especially if the diffusion is not isotropic. Therefore, in this case, the NEMD efficiency calculation is minimized. However, it is necessary even if the field is small enough.

We additionally reviewed the conductivity of $Li_5PS_4Cl_2$ and Li_6PS_5Cl from past and recent works, and also the effect of disarray and defects on diffusion of the Li-ions in Li_6PS_5Cl. Predictions made proved that $Li_5PS_4Cl_2$ is a highly conductive solid electrolyte while its synthesis is in progress. For Li_6PS_5Cl, though the pure material has a moderately low ionic conductivity and was found to be metastable [38,43] (6×10^{-5}–3×10^{-4} S·cm^{-1} at 300 K), it is proven that by intensifying lithium vacancies of the structure and establishing disarray in the ionic positioning of the chlorine and sulfur ions, the ionic conductivity of the material can be improved. Whilst they pronounced those structures formerly, the dimensions of the error bars made it tough to envision the effects. In summary, the experimentally reported increase in conductivity of Li_6PS_5Cl is assumed to be as a result of a combination of lithium-ion vacancies with chlorine–sulfur ion damage, or an increase in halogen (Cl) concentration after heat treatments. The computational results mentioned in a paper by Baktash et al. predict conductivities that are either greater than the actual values or at the upper end of the range of empirical values [19]. The modeling of the system size, the degree of theory employed in the ab initio MD simulations, and changes in the lattice parameters during diffusion carried out under constant volume circumstances are a few systematic flaws in the calculations that might potentially lead to this. Nonetheless, the existence of grain boundaries, contaminants, chaotic, and non-uniform distribution of chaotic sites in the experimental sample may explain the discrepancy between the actual and calculated results. It is important that the calculated results are duplicatable and show trends due to structural changes indicating ways to tune a material to increase conductivity, which is essential for improving SSE.

Based on findings reported by Zhao et al. [89] about some promising compositions, which can be further verified experimentally, it should be noted that some substitutions may be challenging due to the limited solid solution ranges and may require special synthesis conditions [90]. Simultaneously, the solubility itself can be included in the ML predictive model. This will be the focus of future research work.

From reports from the literature, the creation of mechanical stress during discharge in a dual porous insertion electrode cell sandwich made of lithium cobalt oxide and carbon is simulated using models that are given [91]. Models credit two separate factors—changes in the lattice volume brought on by intercalation and phase transformation during the charge or discharge process—to the stress accumulation within intercalation electrodes. Models are used to forecast how cell design elements such as electrode thickness, porosity, and particle size will affect how much stress is generated. To comprehend the mechanical deterioration in a porous electrode during an intercalation or de-intercalation process, models described in the literature are employed. The usage of these models leads to an improved design for battery electrodes that are mechanically durable over an extended time of operation. Refs. [92–103] are recommended to readers for these models.

9. Conclusions

The lithium conductivity in argyrodite electrolytes makes ASSB transcend to be a preferable option for clean electrical energy storage. Scholars have considered a scalable solution-synthetic approach to model argyrodite phases using solvate complexes to enhance elevated ionic conductivities of about 3.9 mS·cm^{-1} with minimal electronic conductivities. ASSB showed great stability performance and good agreement between cells formed by ball-milling and solution-engineered approaches.

Researchers have shown that the AI-NEMD simulation provides a good approximation of the coefficient of diffusivity found in materials with diffusivity that are not intended to be determined directly by the AI-EMD computations. They calculated diffusion coefficients using AI–NEMD and AI–EMD simulations for highly diffusive SSE obtaining accuracy. It was proven from the literature that the AI–EMD approach tends to be slow when used to compute the diffusion coefficients, whereas it is better when using AI–NEMD simulations.

Using AI–NEMD simulations, Li_6PS_5Cl and $Li_5PS_4Cl_2$ were separated as two potential electrolytes for ASSB. It was demonstrated that by including Li vacancies and disorder into the argyrodite electrolyte system, the ionic conductivity of that material will improve substantially.

We reviewed sulfide-based electrolytes reported in the literature as a few potential materials of ASSB because of their mechanical qualities, ionic conductivities, and electrochemical stability surpassing many other potential SSE. We reported that it is possible to construct structures that are stable in the higher conductivity phase at room temperature by introducing halogens into argyrodite models, such as Li_7PS_6.

It was proven that ionic conductivity is good for heterogeneous Cl and Br phase and Cl-rich phases, yielding about 4 mS·cm^{-1}. This shows that the dopants impurities effects of the synthesized argyrodites electrolyte materials does not reduce Li-ions conductivity in the exceptionally fine electrolytes.

Certainly, so far, cells in the literature have been reported to utilize lithium alloy as an anodic electrode due to its toughness, with the goal of measuring the qualities of argyrodite. Future research is expected to make use of lithium as a discrete cathodic substance to see its reaction.

Additionally, future work is expected to factor the ML predictive model extensively into the synthesis of argyrodite electrolyte materials.

Overall, we plan to make this article an all-in-one resource for researchers to access information on the previous and current work on the modeling of argyrodite electrolyte materials with their related references. We included models and methods used by researchers to model the argyrodite electrolyte materials they considered. We also introduced other resources to access more models of argyrodite using ML.

Author Contributions: Conceptualization, S.F., A.B., O.M.A. and S.A.O.; Supervision, S.F. and A.B.; Writing—original draft, O.M.A.; Writing—review & editing, S.F., A.B., S.A.O. All authors have read and agreed to the published version of the manuscript.

Funding: This research received no external funding.

Informed Consent Statement: Not applicable.

Acknowledgments: The support of the University of Akron is highly appreciated.

Conflicts of Interest: The authors declare no conflict of interest.

Nomenclature

Symbol	Description
E_a	activation energy
n	band index
k_B	Boltzmann's constant
W_k	Brillioun zone sampling factor
D_c	Lithium center-of-mass coefficient of diffusion
c_i	color charge/label
J_c	color current
F_c	color field
α	Nosé–Hoover thermostat couples of atoms
D_0	diffusion pre-exponential factor
$\Delta r_i(t)$	ith displacement of N lithium-ions over a period
e	elementary electron charge
$\langle\,\rangle$	ensemble average

Symbol	Description
Q_1, Q_2 and Q_3	friction coefficients
F_i	inter-atomic force on atom i
n	ion density of Li
σ	ionic conductivity of the material
m_i	mass of atom i
p_i	momentum
g	number of degrees of freedom
N	number of lithium-ions subject to a color field
$N^a(E)$	partial densities states
D_0	position
$r_i(t)$	ith position of lithium-ion at time t
D_s	coefficient of self-diffusion
Δ	smearing factor
T	target temperature
Z	valence of Li
v_i	velocity of ith Li-ion
k	wave vector
f^a_{gnk}	weighing factor
Å	Angstrom
AI–EMD/EMD	Ab initio equilibrium molecular dynamics
AI–NEMD/NEMD	Ab initio nonequilibrium molecular dynamics
ASSB	All-solid-state batteries
ASSLB	All-solid-state lithium battery
DC	Direct current
DFT	Density functional theory
DME	Dimethoxy ethane
EC	Electronic conductivity
EIS	Electrochemical impedance spectroscopy
GGA	Generalized gradient approximation
GPW	Gaussian and plane-wave
IC	Ionic conductivity
IL	Ionic liquid
LIB	Lithium-ion batteries
MD	Molecular dynamics
ML	Machine learning
MSD	Mean square displacement
PBE	Perdew–Burke–Ernzerhof
Ry	Rydberg
SEM	Scanning electron microscope
SS	Stainless steel
SSB	Solid-state batteries
SSE	Solid-state electrolyte
THF	Tetrahydrofuran
VASP	Vienna Ab initio Simulation Package
XRD	X-ray diffraction
Symbol	Description

References

1. Wang, Y.; Richards, W.D.; Ong, S.P.; Miara, L.J.; Kim, J.C.; Mo, Y.; Ceder, G. Design principles for solid-state lithium superionic conductors. *Nat. Mater.* **2015**, *14*, 1026–1031. [CrossRef] [PubMed]
2. Strauss, F.; Bartsch, T.; de Biasi, L.; Kim, A.-Y.; Janek, J.; Hartmann, P.; Brezesinski, T. Impact of Cathode Material Particle Size on the Capacity of Bulk-Type All-Solid-State Batteries. *ACS Energy Lett.* **2018**, *3*, 992–996. [CrossRef]
3. Richards, W.D.; Miara, L.J.; Wang, Y.; Kim, J.C.; Ceder, G. Interface Stability in Solid-State Batteries. *Chem. Mater.* **2015**, *28*, 266–273. [CrossRef]
4. Zhang, Z.; Shao, Y.; Lotsch, B.; Hu, Y.-S.; Li, H.; Janek, J.; Nazar, L.F.; Nan, C.-W.; Maier, J.; Armand, M.; et al. New horizons for inorganic solid state ion conductors. *Energy Environ. Sci.* **2018**, *11*, 1945–1976. [CrossRef]
5. Manthiram, A.; Yu, X.; Wang, S. Lithium battery chemistries enabled by solid-state electrolytes. *Nat. Rev. Mater.* **2017**, *2*, 16103. [CrossRef]

6. Xin, S.; You, Y.; Wang, S.; Gao, H.-C.; Yin, Y.-X.; Guo, Y.-G. Solid-State Lithium Metal Batteries Promoted by Nanotechnology: Progress and Prospects. *ACS Energy Lett.* **2017**, *2*, 1385–1394. [CrossRef]

7. Yu, C.; van Eijck, L.; Ganapathy, S.; Wagemaker, M. Synthesis, structure and electrochemical performance of the argyrodite Li_6PS_5Cl solid electrolyte for Li-ion solid state batteries. *Electrochim. Acta* **2016**, *215*, 93–99. [CrossRef]

8. Zhou, L.; Park, K.H.; Sun, X.; Lalère, F.; Adermann, T.; Hartmann, P.; Nazar, L.F. Solvent-Engineered Design of Argyrodite Li_6PS_5X (X = Cl, Br, I) Solid Electrolytes with High Ionic Conductivity. *ACS Energy Lett.* **2019**, *4*, 265–270. [CrossRef]

9. Zhou, W.; Hao, F.; Fang, D. The effects of elastic stiffening on the evolution of the stress field within a spherical electrode particle of lithium-ion batteries. *Int. J. Appl. Mech.* **2013**, *5*, 1350040. [CrossRef]

10. Wang, Z.; Santhanagopalan, D.; Zhang, W.; Wang, F.; Xin, H.L.; He, K.; Li, J.; Dudney, N.; Meng, Y.S. In Situ STEM-EELS Observation of Nanoscale Interfacial Phenomena in All-Solid-State Batteries. *Nano Lett.* **2016**, *16*, 3760–3767. [CrossRef]

11. Zhang, W.; Weber, D.A.; Weigand, H.; Arlt, T.; Manke, I.; Schröder, D.; Koerver, R.; Leichtweiss, T.; Hartmann, P.; Zeier, W.G.; et al. Interfacial Processes and Influence of Composite Cathode Microstructure Controlling the Performance of All-Solid-State Lithium Batteries. *ACS Appl. Mater. Interfaces* **2017**, *9*, 17835–17845. [CrossRef] [PubMed]

12. Wang, S.; Xu, H.; Li, W.; Dolocan, A.; Manthiram, A. Interfacial Chemistry in Solid-State Batteries: Formation of Interphase and Its Consequences. *J. Am. Chem. Soc.* **2017**, *140*, 250–257. [CrossRef] [PubMed]

13. Wenzel, S.; Leichtweiss, T.; Krüger, D.; Sann, J.; Janek, J. Interphase formation on lithium solid electrolytes—An in situ approach to study interfacial reactions by photoelectron spectroscopy. *Solid State Ion.* **2015**, *278*, 98–105. [CrossRef]

14. Seo, D.-H.; Lee, J.; Urban, A.; Malik, R.; Kang, S.; Ceder, G. The structural and chemical origin of the oxygen redox activity in layered and cation-disordered Li-excess cathode materials. *Nat. Chem.* **2016**, *8*, 692–697. [CrossRef] [PubMed]

15. He, X.; Zhu, Y.; Epstein, A.; Mo, Y. Statistical variances of diffusional properties from ab initio molecular dynamics simulations. *NPJ Comput. Mater.* **2018**, *4*, 18. [CrossRef]

16. Baktash, A.; Reid, J.C.; Yuan, Q.; Roman, T.; Searles, D.J. Shaping the Future of Solid-State Electrolytes through Computational Modeling. *Adv. Mater.* **2020**, *32*, e1908041. [CrossRef]

17. Deng, Z.; Zhu, Z.; Chu, I.-H.; Ong, S.P. Data-Driven First-Principles Methods for the Study and Design of Alkali Superionic Conductors. *Chem. Mater.* **2016**, *29*, 281–288. [CrossRef]

18. Aeberhard, P.C.; Williams, S.R.; Evans, D.J.; Refson, K.; David, W.I. Ab initio Nonequilibrium Molecular Dynamics in the Solid Superionic Conductor LiBH4. *Phys. Rev. Lett.* **2012**, *108*, 095901. [CrossRef]

19. Baktash, A.; Reid, J.C.; Roman, T.; Searles, D.J. Diffusion of lithium ions in Lithium-argyrodite solid-state. *NPJ Comput. Mater.* **2020**, *6*, 162. [CrossRef]

20. Rayavarapu, P.R.; Sharma, N.; Peterson, V.K.; Adams, S. Variation in structure and Li+-ion migration in argyrodite-type Li_6PS_5X (X = Cl, Br, I) solid electrolytes. *J. Solid State Electrochem.* **2012**, *16*, 1807–1813. [CrossRef]

21. De Klerk, N.J.; Rosłoń, I.; Wagemaker, M. Diffusion Mechanism of Li Argyrodite Solid Electrolytes for Li-ion Batteries and Prediction of Optimized Halogen Doping: The Effect of Li Vacancies, Halogens, and Halogen Disorder. *Chem. Mater.* **2016**, *28*, 7955–7963. [CrossRef]

22. Sun, X.; Sun, Y.; Cao, F.; Li, X.; Sun, S.; Liu, T.; Wu, J. Preparation, characterization and ionic conductivity studies of composite sulfide solid electrolyte. *J. Alloy. Compd.* **2017**, *727*, 1136–1141. [CrossRef]

23. Park, K.H.; Oh, D.Y.; Choi, Y.E.; Nam, Y.J.; Han, L.; Kim, J.Y.; Xin, H.; Lin, F.; Oh, S.M.; Jung, Y.S. Solution-Processable Glass LiI-Li4SnS4 Superionic Conductors for All-Solid-State Li-Ion Batteries. *Adv. Mater.* **2015**, *28*, 1874–1883. [CrossRef] [PubMed]

24. Wu, X.; El Kazzi, M.; Villevieille, C. Surface and morphological investigation of the electrode/electrolyte properties in an all-solid-state battery using a Li_2S-P_2S_5 solid electrolyte. *J. Electroceram.* **2017**, *38*, 207–214. [CrossRef]

25. Xu, R.; Wang, X.; Zhang, S.; Xia, Y.; Xia, X.; Wu, J.; Tu, J. Rational coating of $Li_7P_3S_{11}$ solid electrolyte on MoS_2 electrode for all-solid-state lithium ion batteries. *J. Power Sources* **2017**, *374*, 107–112. [CrossRef]

26. Wang, Z.; Shao, G. Theoretical design of solid electrolytes with superb ionic conductivity: Alloying effect on Li+ transportation in cubic Li_6PA_5X chalcogenides. *J. Mater. Chem. A* **2017**, *5*, 1846–21857. [CrossRef]

27. Deiseroth, H.-J.; Maier, J.; Weichert, K.; Nickel, V.; Kong, S.-T.; Reiner, C. Li7PS6 and Li_6PS_5X (X: Cl, Br, I): Possible Three-dimensional Diffusion Pathways for Lithium Ions and Temperature Dependence of the Ionic Conductivity by Impedance Measurements. *Z. Für Anorg. Und Allg. Chem.* **2011**, *637*, 1287–1294. [CrossRef]

28. Kraft, M.A.; Culver, S.P.; Calderon, M.; Böcher, F.; Krauskopf, T.; Senyshyn, A.; Dietrich, C.; Zevalkink, A.; Janek, J.; Zeier, W.G. Influence of Lattice Polarizability on the Ionic Conductivity in the Lithium Superionic Argyrodites Li_6PS_5X (X = Cl, Br, I). *J. Am. Chem. Soc.* **2017**, *139*, 10909–10918. [CrossRef]

29. Epp, V.; Gün, O.; Deiseroth, H.J.; Wilkening, M. Highly Mobile Ions: Low-Temperature NMR Directly Probes Extremely Fast Li+ Hopping in Argyrodite-Type Li_6PS_5Br. *J. Phys. Chem.* **2013**, *4*, 2118–2123. [CrossRef]

30. Rao, R.P.; Sharma, N.; Peterson, V.; Adams, S. Formation and conductivity studies of lithium argyrodite solid electrolytes using in-situ neutron diffraction. *Solid State Ion.* **2013**, *230*, 72–76. [CrossRef]

31. Boulineau, S.; Courty, M.; Tarascon, J.-M.; Viallet, V. Mechanochemical synthesis of Li-argyrodite Li_6PS_5X (X = Cl, Br, I) as sulfur-based solid electrolytes for all solid state batteries application. *Solid State Ion.* **2012**, *221*, 1–5. [CrossRef]

32. Yubuchi, S.; Teragawa, S.; Aso, K.; Tadanaga, K.; Hayashi, A.; Tatsumisago, M. Preparation of High Lithium-ion Conducting Li_6PS_5Cl Solid Electrolyte from Ethanol Solution for All-Solid-State Preparation of High Lithium-ion Conducting. *J. Power Sources* **2015**, *293*, 941–945. [CrossRef]

33. Knauth, P. Inorganic solid Li ion conductors: An overview. *Solid State Ion.* **2009**, *180*, 911–916. [CrossRef]
34. Culver, S.P.; Koerver, R.; Krauskopf, T.; Zeier, W.G. Designing Ionic Conductors: The Interplay between Structural Phenomena and Interfaces in Thiophosphate-Based Solid-State Batteries. *Chem. Mater.* **2018**, *30*, 4179–4192. [CrossRef]
35. Sakuda, A.; Hayashi, A.; Tatsumisago, M. Sulfide Solid Electrolyte with Favorable Mechanical Property for All-Solid-State Lithium Battery. *Sci. Rep.* **2013**, *3*, 2261. [CrossRef]
36. Ni, J.E.; Case, E.D.; Sakamoto, J.S.; Rangasamy, E.; Wolfenstine, J.B. Room temperature elastic moduli and Vickers hardness of hot-pressed LLZO cubic garnet. *J. Mater. Sci.* **2012**, *47*, 7978–7985. [CrossRef]
37. Kanno, R.; Murayama, M. Lithium Ionic Conductor Thio-LISICON: The Li_2S-GeS_2-P_2S_5 System. *J. Electrochem. Soc.* **2001**, *148*, A742–A746. [CrossRef]
38. Kamaya, N.; Homma, K.; Yamakawa, Y.; Hirayama, M.; Kanno, R.; Yonemura, M.; Kamiyama, T.; Kato, Y.; Hama, S.; Kawamoto, K.; et al. A lithium superionic conductor. *Nat. Mater.* **2011**, *10*, 682–686. [CrossRef]
39. Kuhn, A.; Gerbig, O.; Zhu, C.; Falkenberg, F.; Maier, J.; Lotsch, B.V. A New Ultrafast Superionic Li-Conductor: Ion Dynamics in $Li_{11}Si_2PS_{12}$ and Comparison with other Tetragonal LGPS-type Electrolytes. *Phys. Chem. Chem. Phys.* **2014**, *16*, 14669–14674. [CrossRef]
40. Bron, P.; Johansson, S.; Zick, K.; Schmedt auf der Günne, J.; Dehnen, S.; Roling, B. $Li_{10}SnP_2S_{12}$: An Affordable Lithium Superionic Conductor. *J. Am. Chem. Soc.* **2013**, *135*, 15694–15697. [CrossRef]
41. Kato, Y.; Hori, S.; Saito, T.; Suzuki, K.; Hirayama, M.; Mitsui, A.; Yonemura, M.; Iba, H.; Kanno, R. High-power all-solid-state batteries using sulfide superionic conductors. *Nat. Energy* **2016**, *1*, 16030. [CrossRef]
42. Yamane, H.; Shibata, M.; Shimane, Y.; Junke, T.; Seino, Y.; Adams, S.; Minami, K.; Hayashi, A.; Tatsumisago, M. Crystal Structure of a Superionic Conductor, $Li_7P_3S_{11}$. *Solid State Ion.* **2007**, *178*, 1163–1167. [CrossRef]
43. Seino, Y.; Ota, T.; Takada, K.; Hayashi, A.; Tatsumisago, M. A sulphide lithium super ion conductor is superior to liquid ion conductors for use in rechargeable batteries. *Energy Environ. Sci.* **2013**, *7*, 627–631. [CrossRef]
44. Chu, I.-H.; Nguyen, H.; Hy, S.; Lin, Y.-C.; Wang, Z.; Xu, Z.; Deng, Z.; Meng, Y.S.; Ong, S.P. Correction to Insights into the Performance Limits of the $Li_7P_3S_{11}$ Superionic Conductor: A Combined First-Principles and Experimental Study. *ACS Appl. Mater. Interfaces* **2018**, *10*, 7843–7853. [CrossRef]
45. Deiseroth, H.-J.; Kong, S.-T.; Eckert, H.; Vannahme, J.; Reiner, C.; Zaiß, T.; Schlosser, M. Li_6PS_5X: A Class of Crystalline Li-Rich Solids with an Unusually High Li+ Mobility. *Angew. Chem. Int. Ed.* **2008**, *47*, 755–758. [CrossRef]
46. Yu, C.; Ganapathy, S.; Van Eck, E.R.H.; Wang, H.; Basak, S.; Li, Z.; Wagemaker, M. Accessing the Bottleneck in All-Solid State Batteries, Lithium-Ion Transport over the Solid-Electrolyte-Electrode Interface. *Nat. Commun.* **2017**, *8*, 1086–1094. [CrossRef] [PubMed]
47. Wenzel, S.; Sedlmaier, S.J.; Dietrich, C.; Zeier, W.G.; Janek, J. Interfacial reactivity and interphase growth of argyrodite solid electrolytes at lithium metal electrodes. *Solid State Ion.* **2018**, *318*, 102–112. [CrossRef]
48. Chen, M.; Rao, R.P.; Adams, S. High capacity all-solid-state Cu–Li_2S/Li_6PS_5Br/In batteries. *Solid State Ion.* **2014**, *262*, 183–187. [CrossRef]
49. Chen, M.; Rao, R.P.; Adams, S. The unusual role of Li_6PS_5Br in all-solid-state CuS/Li_6PS_5Br/In–Li batteries. *Solid State Ion.* **2014**, *268*, 300–304. [CrossRef]
50. Chen, M.; Yin, X.; Reddy, M.V.; Adams, S. All-Solid-State MoS_2/Li_6PS_5Br/In-Li Batteries as a Novel Type of Li/S Battery. *J. Mater. Chem. A* **2015**, *3*, 10698–10702. [CrossRef]
51. Chen, M.; Adams, S. High performance all-solid-state lithium/sulfur batteries using lithium argyrodite electrolyte. *J. Solid State Electrochem.* **2014**, *19*, 697–702. [CrossRef]
52. Auvergniot, J.; Cassel, A.; Ledeuil, J.B.; Viallet, V.; Seznec, V.; Dedryvère, R. Interface Stability of Argyrodite Li6PS5Cl toward $LiCoO_2$, $LiNi_{1/3}Co_{1/3}Mn_{1/3}O_2$, and $LiMn_2O_4$ in Bulk All-Solid-State Batteries. *Chem. Mater.* **2017**, *29*, 3883–3890. [CrossRef]
53. Yu, C.; Ganapathy, S.; De Klerk, N.J.; Roslon, I.; van Eck, E.R.; Kentgens, A.P.; Wagemaker, M. Unravelling Li-ion Transport from Picoseconds to Seconds: Bulk versus Interfaces in an Argyrodite Li_6PS_5Cl-Li_2S All-Solid-State Li-Ion Battery. *J. Am. Chem. Soc.* **2016**, *138*, 11192–11201. [CrossRef] [PubMed]
54. Kim, D.H.; Oh, D.Y.; Park, K.H.; Choi, Y.E.; Nam, Y.J.; Lee, H.A.; Lee, S.-M.; Jung, Y.S. Infiltration of Solution-Processable Solid Electrolytes into Conventional Li-Ion-Battery Electrodes for All-Solid-State Li-Ion Batteries. *Nano Lett.* **2017**, *17*, 3013–3020. [CrossRef] [PubMed]
55. Sedlmaier, S.J.; Indris, S.; Dietrich, C.; Yavuz, M.; Dräger, C.; von Seggern, F.; Sommer, H.; Janek, J. Li_4PS_4I: A Li+ Superionic Conductor Synthesized by a Solvent-Based Soft Chemistry Approach. *Chem. Mater.* **2017**, *29*, 1830–1835. [CrossRef]
56. Liu, Z.; Fu, W.; Payzant, E.A.; Yu, X.; Wu, Z.; Dudney, N.J.; Kiggans, J.; Hong, K.; Rondinone, A.J.; Liang, C. Anomalous High Ionic Conductivity of Nanoporous β-Li_3PS_4. *J. Am. Chem. Soc.* **2013**, *135*, 975–978. [CrossRef] [PubMed]
57. Rangasamy, E.; Liu, Z.; Gobet, M.; Pilar, K.; Sahu, G.; Zhou, W.; Wu, H.; Greenbaum, S.; Liang, C. An Iodide-based $Li_7P_2S_8I$ Superionic Conductor. *J. Am. Chem. Soc.* **2015**, *137*, 1384–1387. [CrossRef]
58. Ito, S.; Nakakita, M.; Aihara, Y.; Uehara, T.; Machida, N. A Synthesis of Crystalline $Li_7P_3S_{11}$ Solid Electrolyte from 1,2-dimethoxyethane Solvent. *J. Power Sources* **2014**, *271*, 342–345. [CrossRef]
59. Yubuchi, S.; Uematsu, M.; Deguchi, M.; Hayashi, A.; Tatsumisago, M. Lithium-Ion-Conducting Argyrodite-Type Li_6PS_5X (X = Cl, Br, I) Solid Electrolytes Prepared by a Liquid-Phase Technique Using Ethanol as a Solvent. *ACS Appl. Energy Mater.* **2018**, *1*, 3622–3629. [CrossRef]

60. Pecher, O.; Kong, S.T.; Goebel, T.; Nickel, V.; Weichert, K.; Reiner, C.; Deiseroth, H.J.; Maier, J.; Haarmann, F.; Zahn, D. Atomistic Characterisation of Li+ Mobility and Conductivity in $Li_{7-x}PS_{6-x}I_x$ Argyrodites from Molecular Dynamics Simulations, Solid-State NMR, and Impedance Spectroscopy. *Chem.-Eur. J.* **2010**, *16*, 8347–8354. [CrossRef]

61. Evans, D.J.; Morriss, G.P. *Statistical Mechanics of Nonequilibrium Liquids*; ANU E Press: Canberra, Australia, 2007.

62. Zhu, Z.; Chu, I.H.; Deng, Z.; Ong, S.P. Role of Na+ interstitials and dopants in enhancing the Na+ conductivity of the cubic Na3PS4 superionic conductor. *Chem. Mat.* **2015**, *27*, 8318–8325. [CrossRef]

63. Marcolongo, A.; Marzari, N. Ionic correlations and failure of Nernst-Einstein relation in solid-state electrolytes. *Phys. Rev. Mater.* **2017**, *1*, 025402. [CrossRef]

64. Keil, F.J.; Krishna, R.; Coppens, M.-O. Modeling of Diffusion in Zeolites. *Rev. Chem. Eng.* **2000**, *16*, 71–197. [CrossRef]

65. Rosero-Navarro, N.C.; Miura, A.; Tadanaga, K. Preparation of lithium ion conductive Li_6PS_5Cl solid electrolyte from solution for the fabrication of composite cathode of all-solid-state lithium battery. *J. Sol-Gel Sci. Technol.* **2018**, *89*, 303–309. [CrossRef]

66. Yu, C.; Ganapathy, S.; Hageman, J.; van Eijck, L.; van Eck, E.R.H.; Zhang, L.; Schwietert, T.; Basak, S.; Kelder, E.M.; Wagemaker, M. Facile Synthesis toward the Optimal Structure-Conductivity Characteristics of the Argyrodite Li_6PS_5Cl Solid-State Electrolyte. *ACS Appl. Mater. Interfaces* **2018**, *10*, 33296–33306. [CrossRef]

67. Rao, R.P.; Adams, S. Studies of lithium argyrodite solid electrolytes for all-solid-state batteries. *Phys. Status Solidi A* **2011**, *208*, 1804–1807. [CrossRef]

68. Wang, S.; Zhang, Y.; Zhang, X.; Liu, T.; Lin, Y.H.; Shen, Y.; Li, L.; Nan, C.W. High-Conductivity Argyrodite Li_6PS_5Cl Solid Electrolytes Prepared via Optimized Sintering Processes for All-Solid-State Lithium-Sulfur Batteries. *ACS Appl. Mater. Interfaces* **2018**, *10*, 42279–42285. [CrossRef]

69. Zhang, K.; Zhou, G.; Fang, T.; Jiang, K.; Liu, X. Structural Reorganization of Ionic Liquid Electrolyte by a Rapid Charge/Discharge Circle. *J. Phys. Chem. Lett.* **2021**, *12*, 2273–2278. [CrossRef]

70. Lynden-Bell, R.M. Gas–Liquid interfaces of room temperature ionic liquids. *Mol. Phys.* **2003**, *101*, 2625–2633. [CrossRef]

71. Lynden-Bell, R.M.; Kohanoff, J.; Del Popolo, M.G. Simulation of interfaces between room temperature ionic liquids and other liquids. *Faraday Discuss.* **2004**, *129*, 57–67. [CrossRef]

72. Fedorov, M.V.; Kornyshev, A.A. Ionic Liquids at Electrified Interfaces. *Chem. Rev.* **2014**, *114*, 2978–3036. [CrossRef] [PubMed]

73. Hutter, J.; Iannuzzi, M.; Schiffmann, F.; VandeVondele, J. CP2K: Atomistic simulations of condensed matter systems. *Wiley Interdiscip. Rev. Comput. Mol. Sci.* **2014**, *4*, 15–25. [CrossRef]

74. VandeVondele, J.; Krack, M.; Mohamed, F.; Parrinello, M.; Chassaing, T.; Hutter, J. Quickstep: Fast and accurate density functional calculations using a mixed Gaussian and plane waves approach. *Comput. Phys. Commun.* **2005**, *167*, 103–128. [CrossRef]

75. Kresse, G.; Furthmüller, J. Efficient iterative schemes for abinitio total-energy calculations using a plane-wave basis set. *Phys. Rev. B Condens. Matter Mater. Phys.* **1996**, *54*, 11169−11186. [CrossRef] [PubMed]

76. Grimme, S.; Antony, J.; Ehrlich, S.; Krieg, H. A consistent and accurate ab initio parametrization of density functional dispersion correction (DFT-D) for the 94 elements H-Pu. *J. Chem. Phys.* **2010**, *132*, 154104–154119. [CrossRef] [PubMed]

77. Goedecker, S.; Teter, M.; Hutter, J. Separable dual-space Gaussian pseudopotentials. *Phys. Rev. B* **1996**, *54*, 1703–1710. [CrossRef] [PubMed]

78. VandeVondele, J.; Hutter, J. Gaussian basis sets for accurate calculations on molecular systems in gas and condensed phases. *J. Chem. Phys.* **2007**, *127*, 114105. [CrossRef]

79. Lippert, G.; Hutter, J.; Parrinello, M. A hybrid Gaussian and plane wave density functional scheme. *Mol. Phys.* **1997**, *92*, 477–487. [CrossRef]

80. Evans, D.J.; Morriss, O. Non-Newtonian molecular dynamics. *Comput. Phys. Rep.* **1984**, *1*, 297–343. [CrossRef]

81. Wheeler, D.R.; Newman, J. Molecular dynamics simulations of multicomponent diffusion. 2. Nonequilibrium method. *J. Phys. Chem. B* **2004**, *108*, 18362–18367. [CrossRef]

82. Arya, G.; Chang, H.-C.; Maginn, E. A critical comparison of equilibrium, non-equilibrium and boundary-driven molecular dynamics techniques for studying transport in microporous materials. *J. Chem. Phys.* **2001**, *115*, 8112–8124. [CrossRef]

83. Brańka, A.C. Nosé-Hoover chain method for nonequilibrium molecular dynamics simulation. *Phys. Rev. E* **2000**, *61*, 4769–4773. [CrossRef] [PubMed]

84. Beedle, A.E.M.; Mora, M.; Davis, C.T.; Snijders, A.P.; Stirnemann, G.; Garcia-Manyes, S. Forcing the reversibility of a mechanochemical reaction. *Nat. Commun.* **2018**, *9*, 3155. [CrossRef] [PubMed]

85. Holzwarth, N.; Lepley, N.; Du, Y. A Computer Modeling Study of Lithium Phosphate and Thiophosphate Electrolyte Materials. *ECS Meet. Abstr.* **2010**, *1*, 517. [CrossRef]

86. Grasselli, F.; Baroni, S. Topological quantization and gauge invariance of charge transport in liquid insulators. *Nat. Phys.* **2019**, *15*, 967–972. [CrossRef]

87. Electronic Design. 2020. Available online: https://www.electronicdesign.com/markets/article/21210395/machine-learning-boosts-electrolyte-search (accessed on 21 August 2022).

88. Gallo-Bueno, A.; Reynaud, M.; Casas-Cabanas, M.; Carrasco, J. Unsupervised machine learning to classify crystal structures according to their structural distortion: A case study on Li-argyrodite solid-state electrolytes. *Energy AI* **2022**, *9*, 100159. [CrossRef]

89. Zhao, Q.; Avdeev, M.; Chen, L.; Shi, S. Machine learning prediction of activation energy in cubic Li-argyrodites with hierarchically encoding crystal structure-based (HECS) descriptors. *Sci. Bull.* **2021**, *66*, 1401–1408. [CrossRef]

90. Zhang, Z.; Sun, Y.; Duan, X.; Peng, L.; Jia, H.; Zhang, Y.; Shan, B.; Xie, J. Design and synthesis of room temperature stable Li-argyrodite superionic conductors via cation doping. *J. Mater. Chem. A* **2019**, *7*, 2717–2722. [CrossRef]
91. Christensen, J.; Newman, J. A Mathematical Model of Stress Generation and Fracture in Lithium Manganese Oxide. *J. Electrochem. Soc.* **2006**, *153*, A1019. [CrossRef]
92. Zhou, W. Effects of external mechanical loading on stress generation during lithiation in Li-ion battery electrodes. *Electrochim. Acta* **2015**, *185*, 28–33. [CrossRef]
93. Golmon, S.; Maute, K.; Lee, S.-H.; Dunn, M.L. Stress generation in silicon particles during lithium insertion. *Appl. Phys. Lett.* **2010**, *97*, 033111. [CrossRef]
94. Wan, X.; Zhang, Z.; Niu, H.; Yin, Y.; Kuai, C.; Wang, J.; Shao, C.; Guo, Y. Machine-Learning-Accelerated Catalytic Activity Predictions of Transition Metal Phthalocyanine Dual-Metal-Site Catalysts for CO_2 Reduction. *J. Phys. Chem. Lett.* **2021**, *12*, 6111–6118. [CrossRef] [PubMed]
95. Xiao, R. Modeling Mismatch Strain Induced Self-Folding of Bilayer Gel Structures. *Int. J. Appl. Mech.* **2016**, *8*, 1640004. [CrossRef]
96. Huang, R.; Zheng, S.; Liu, Z.; Ng, T.Y. Recent Advances of the Constitutive Models of Smart Materials—Hydrogels and Shape Memory Polymers. *Int. J. Appl. Mech.* **2020**, *12*, 2050014. [CrossRef]
97. Brassart, L.; Suo, Z. Reactive flow in large-deformation electrodes of lithium-ion batteries. *Int. J. Appl. Mech.* **2012**, *4*, 1250023. [CrossRef]
98. Bhandakkar, T.K.; Gao, H. Cohesive modeling of crack nucleation under diffusion induced stresses in a thin strip: Implications on the critical size for flaw tolerant battery electrodes. *Int. J. Solids Struct.* **2010**, *47*, 1424–1434. [CrossRef]
99. Christensen, J. Modeling Diffusion-Induced Stress in Li-Ion Cells with Porous Electrodes. *J. Electrochem. Soc.* **2010**, *157*, A366–A380. [CrossRef]
100. Cheng, Y.-T.; Verbrugge, M.W. Diffusion-Induced Stress, Interfacial Charge Transfer, and Criteria for Avoiding Crack Initiation of Electrode Particles. *J. Electrochem. Soc.* **2010**, *157*, A508–A516. [CrossRef]
101. Deshpande, R.; Cheng, Y.-T.; Verbrugge, M.W. Modeling diffusion-induced stress in nanowire electrode structures. *J. Power Sources* **2010**, *195*, 5081–5088. [CrossRef]
102. Garcia, R.E.; Chiang, Y.-M.; Carter, W.C.; Limthongkul, P.; Bishop, C. Microstructural Modeling and Design of Rechargeable Lithium-Ion Batteries. *J. Electrochem. Soc.* **2005**, *152*, A255–A263. [CrossRef]
103. Renganathan, S.; Sikha, G.; Santhanagopalan, S.; White, R.E. Theoretical Analysis of Stresses in a Lithium Ion Cell. *J. Electrochem. Soc.* **2010**, *157*, A155. [CrossRef]

Article

Modeling the Effect of Cell Variation on the Performance of a Lithium-Ion Battery Module

Dongcheul Lee, Seohee Kang and Chee Burm Shin *

Department of Chemical Engineering and Division of Energy Systems Research, Ajou University,
Suwon 16499, Korea
* Correspondence: cbshin@ajou.ac.kr

Abstract: Owing to the variation between lithium-ion battery (LIB) cells, early discharge termination and overdischarge can occur when cells are coupled in series or parallel, thereby triggering a decrease in LIB module performance and safety. This study provides a modeling approach that considers the effect of cell variation on the performance of LIB modules in energy storage applications for improving the reliability of the power quality of energy storage devices and efficiency of the energy system. Ohm's law and the law of conservation of charge were employed as the governing equations to estimate the discharge behavior of a single strand composing of two LIB cells connected in parallel based on the polarization properties of the electrode. Using the modeling parameters of a single strand, the particle swarm optimization algorithm was adopted to predict the discharge capacity and internal resistance distribution of 14 strands connected in series. Based on the model of the LIB strand to predict the discharge behavior, the effect of cell variation on the deviation of the discharge termination voltage and depth of discharge imbalance was modeled. The validity of the model was confirmed by comparing the experimental data with the modeling results.

Keywords: lithium-ion battery; modeling; voltage deviation; electrical behaviors; cell performance deviation

Citation: Lee, D.; Kang, S.; Shin, C.B. Modeling the Effect of Cell Variation on the Performance of a Lithium-Ion Battery Module. *Energies* **2022**, *15*, 8054. https://doi.org/10.3390/en15218054

Academic Editor: Siamak Farhad

Received: 7 October 2022
Accepted: 26 October 2022
Published: 29 October 2022

Publisher's Note: MDPI stays neutral with regard to jurisdictional claims in published maps and institutional affiliations.

1. Introduction

Lithium-ion batteries (LIBs) have diverse applications because of their high energy density, efficiency, and power [1]. LIB-powered energy storage devices balance the power supply and demand, reduce peaks, and maintain a stable power system [2]. Cell-to-cell variation is unavoidable due to the inconsistency from the fabrication of cell components including electrodes, separators, and electrolytes to the assembly process of cell [3–7]. Owing to an imbalance between cells, overcharging and overdischarging can occur when LIBs are connected in series or parallel, thereby triggering a decrease in module performance and safety [8]. Therefore, computing the cell-to-cell deviation effect of the LIB module via mathematical modeling is necessary to improve the reliability of the power quality of the energy storage device and ensure efficient operation of the energy system [9].

Numerous studies have reviewed the effects of cell-to-cell deviation in modules and packs of LIBs [3–7]. Xie et al. [10] proposed an experimental procedure to measure the capacity and voltage deviation accurately between cells connected in series of an LIB module. Chang et al. [11] modeled the performance of a battery pack based on Monte Carlo experiments in which LIB cells were connected in parallel. Astaneh et al. [12] predicted the performance of an LIB pack for EV by scale-up, considering the random variability of the cell based on the electro-thermal model of the LIB cell. Tran et al. [13] estimated the state of charge (SOC) distributions of modules connected in series with LIB cells using a nonlinear state observer. They lowered the computational burden compared with the conventional SOC estimation algorithm. Liu and Zhang [14] proposed a cell balancing method to solve the cell imbalance of batteries by designing an active balancing circuit through SOC estimation using an extended Kalman filter. Lee et al. [15] predicted the voltage behavior in a LIB pack based on the normal distribution of the capacity and internal resistance of the LIB cell. They verified the pack-modelling results through a dynamic test. Krupp et al. [16] proposed a modeling tool that estimates the state of health (SOH) distribution of individual cells and that of a module in real-time through incremental capacity analysis. Zilberman et al. [17] estimated the distribution of the capacity and impedance of 18,650 LIB cells using differential voltage analysis and analyzed the deviation changes according to calendar aging. The cell-to-cell deviation that occurs when the battery module is operating can cause overcharge and overdischarge in the cell and accelerate cell aging, reducing the efficiency of the battery module and causing safety issues [18]. Commercial battery modules require a modeling methodology to accurately estimate the cell deviation in real time. The voltage deviation among cells should be predicted by estimating the individual parameters of the battery cell to minimize energy loss in the battery system [19].

Many works mentioned in the above [10–17] reported the estimation of the distributions of the cell voltage, SOC, capacity, and internal resistance distributions of individual cells in LIB modules using various modeling methodologies including extended Kalman filter, nonlinear state observer, incremental capacity analysis, and differential voltage analysis among others. To the best of authors' knowledge, there are no published articles on the assessment of overdischarge and early discharge termination of individual cells in a module based on modeling and verified by the comparison of experimental measurement data. Herein, we propose a methodology that can effectively model the deviation of the discharge termination and depth of discharge (DOD) imbalance of the LIB module under the effect of cell variation. Two cells were combined in parallel to form a single strand, and then 14 strands were connected in series in the LIB module analyzed in this study. Based on the polarization properties of the electrode, Ohm's law and the law of charge conservation were employed as governing equations to predict the discharge performance of a single strand. The distribution of capacity and internal resistance of the LIB strand, which are the main parameters of strand deviation in the LIB module [3,17], were calculated using the particle swarm algorithm (PSO) [20,21]. In the modeling approach proposed in this study, several constant-current discharge experiments of the module were performed to calculate the modeling parameters. Finally, the experimental data were compared with the modeling results to validate the proposed model.

2. Mathematical Model

Herein, an LIB cell by LG Energy Solution with a capacity of 63 Ah composed of lithium nickel manganese cobalt oxide (NMC) cathodes and graphite anodes was modeled. The modeling procedure for each LIB cell in the module was similar to that used in our previous studies [22–24]. As LIB batteries are composed of alternating anodes and cathodes, two parallel cathodes and anodes with tabs positioned in opposite directions, as shown in Figure 1, are modeled. Figure 1 shows that the current flow between the electrodes from the cathode to the anode during discharge was perpendicular to the electrode, assuming that the distance between the cathode and the anode was very close. From the continuity of the current at the cathode and anode during discharge, the Poisson equations for the voltages at the cathode and anode can be derived as follows:

$$\nabla^2 V_c = +r_c J \quad \text{in } \Omega_c \tag{1}$$

$$\nabla^2 V_a = -r_a J \quad \text{in } \Omega_a \tag{2}$$

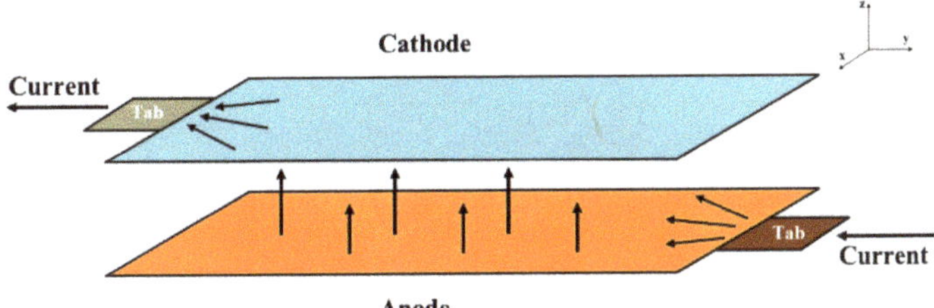

Figure 1. Schematic of the current flow in parallel cathode and anode of LIB during discharge.

Resistances r_c and r_a were calculated as described in the previous papers [22–24]. In addition, a previous study reported relevant boundary conditions for V_c and V_a [22].

Current density, J, is a function of the cathode and anode voltages, and the functional relationship between the current density and voltage depends on the polarization properties of the electrodes. This study adopted the polarization characteristics proposed by Tiedemann and Newman [25] and Newman and Tiedemann [26], as follows:

$$J_{faradaic} = Y(V_c - V_a - U) \tag{3}$$

where Y and U are the fitting parameters as functions of the DOD, as shown in Gu [27], and their electrochemical meanings have been described in the previous studies [28,29]. The detailed procedure for deriving Y and U from the experimental data and the functional relationship between Y and U according to DOD are described in Section 4.

By excluding the modeling procedure to calculate the voltage distribution of the transport phenomena of ionic species and electrolyte phase, the model reduces the computational burden significantly while maintaining its validity [22–24,28–35]. In the simplified model, the DOD according to the time during discharge can be calculated by integrating J with respect to time as follows:

$$DOD = DOD_i + \frac{\int_0^t J d\tau}{Q_0} \tag{4}$$

Figure 2 shows a schematic of an LIB module in which two LIB cells are connected in parallel to form a single strand and 14 strands are connected in series. The negative and positive terminals of the module composed of copper busbars were connected to the 1st and 14th strands, respectively. Each cell was connected in series and parallel with aluminum bus bars through laser welding to minimize the electrical contact resistance, while the 7th and 8th strands were connected with copper busbars. The LIB module in Figure 2 measures the total terminal voltage and voltage at each strand terminals through a sensor. The electrical resistance, including the busbar, BMS sensor, and contact resistance with the terminal, is negligible because of the good welding conditions and very small compared to the resistance of the LIB cell [36]. As the measurement data that can be obtained from the experiment are the terminal voltage of the module and the voltage of 14 strands, we proceeded with the above modeling procedure by considering the strand in which two cells are connected in parallel as a single battery model. The key parameters of the above-mentioned modeling for calculating the voltage deviation of the battery strands are Q_n and Y_n. Q_n is the capacity of the nth LIB strand, and Y_n correspnds to the reciprocal of the internal resistance of the LIB [29]. As Y_n is a function of the DOD, Equation (5) is introduced to efficiently calculate the distribution of Y_n by multiplying Y_{ref} with a proportionality constant, β_n.

$$Y_n = Y_{ref}\beta_n \tag{5}$$

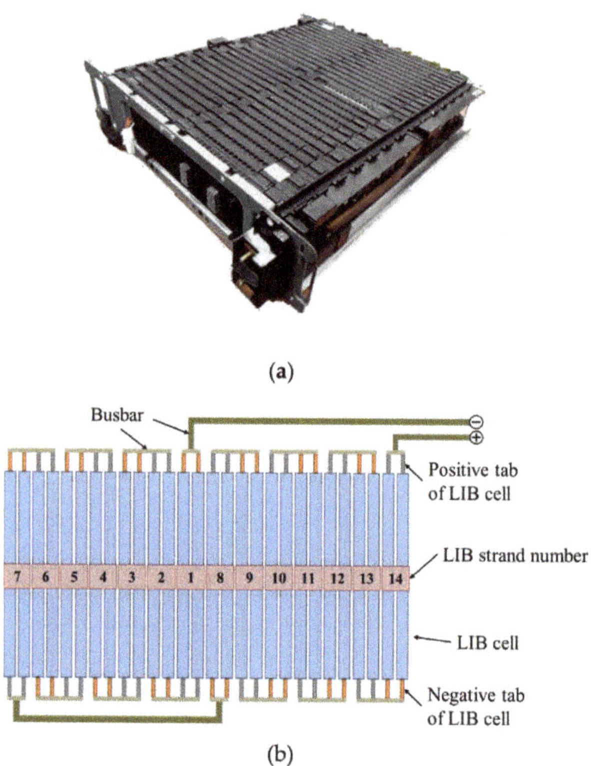

(a)

(b)

Figure 2. (**a**) Photograph of an LIB module with LIB cells configured in two parallel and 14 series and (**b**) schematic of the electrical connections of the LIB module.

To reflect the deviation of Q_n and Y_n from the discharge behavior of the reference LIB strand, the *DOD* of each LIB strand was modified with Q_n and was calculated as follows:

$$DOD_n = DOD_{n,i} + \frac{\int_0^t J d\tau}{Q_n} \tag{6}$$

The PSO algorithm was used to calculate the distribution of strand capacity and internal resistance. PSO is a widely used metaheuristic optimization method that determines an optimal solution by moving particles, a solution candidate group, according to a mathematical formula to solve the problem [20]. The PSO algorithm has fewer hyperparameters than other optimization techniques, and has the advantages of concise theory, efficient operation, and stable convergence [21]. To determine Q_n and β_n using the PSO algorithm, V_{model} according to a constant current was calculated by assigning Q_n and β_n into the battery strand model as inputs. Equation (7) was used as the fitness function to minimize the sum of errors compared with the experimental V_{exp}.

$$fitness\ function = \int_0^t \sqrt{\left(V_{exp} - V_{model}\right)^2} d\tau \tag{7}$$

3. Experimental Section

The energy of the LIB cell composed of the NMC cathode and graphite anode was 232 Wh, and the nominal capacity was 63 Ah. The positive and negative tabs of the LIB cells were positioned in both directions. The energy of the two parallel and 14 series-connected LIB modules is 6.5 kWh, and the nominal capacity is 126 Ah. The LIB module was stacked with an aluminum case of dimensions $445 \times 110 \times 620$ mm^3. In Figure 2, the negative tab of the 1st strand is connected to the negative terminal of the module. The positive tab of the 14th strand is connected to the positive terminal of the module. LIB module tests were conducted in a chamber maintained at 25 °C. The LIB module was charged with a current of 0.2 C and discharged at three different C rates of 0.2 C, 0.5 C, and 1 C with a rest of 1 h in the same experimental condition as the cell test. The discharge termination voltage of the LIB cell was 42 V. During the experiment, the voltage sensor measured the terminal voltage of the module and those of the 14 strands.

4. Results and Discussion

4.1. Model Validation for LIB Strand

As one of the experimental data that can be obtained from the discharge test of the LIB module is the voltage of a strand, the LIB strand was modeled as a single-battery model by neglecting the electrical resistance for parallel connection in this work. The nominal capacity of the LIB strand was 126 Ah, because the two LIB cells were connected in parallel. Figure 3 shows the discharge voltage according to three different C rates in the range of 0.2 C to 1 C of the 1st strand. The discharge curves in Figure 3 show the nonlinear behaviors, because the electrical resistance of battery cell is not constant during the discharge with constant current as shown by combinatorial atomistic-to-AI simulation on supercapacitors validated by experimental data [37]. In this work, the internal resistance change of LIB cell during discharge was accounted through the functional dependence of model parameter Y on DOD, because Y can be regarded as the reciprocal of the internal resistance of the LIB [28,29]. The dependence of Y on the environmental temperature was reported in the previous work [28]. Figure 4 illustrates the linear relationship between the LIB strand voltage and current density during discharge obtained from the experimental results shown in Figure 3. The well-fitted results in Figure 4 justify the linear relationship between the current density and LIB strand voltage of Equation (3). As the vertical intercept and slope of the linear function change according to the DOD level, U and Y can be expressed as functions of DOD. Unlike the previous works [22–24,28] in which U and Y were expressed as the polynomial functions of DOD, U and Y were determined through an interpolation with a 2×101 matrix whose elements are the values U and Y at a certain DOD between 0 and 1 with an increment of 0.01 given in Figure 5a,b. U can be considered as the equilibrium voltage and Y as the reciprocal of the internal resistance of the LIB [28,29].

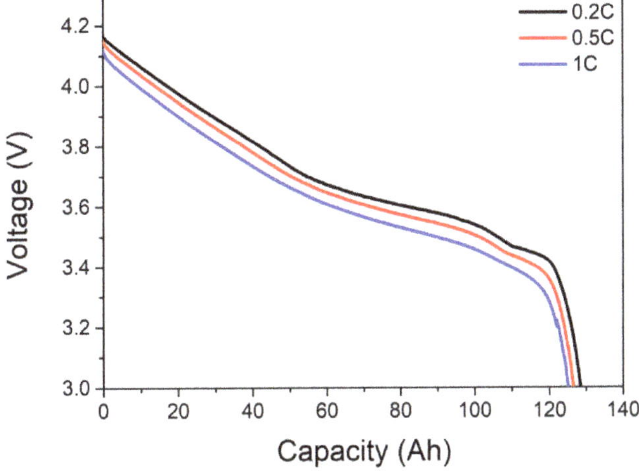

Figure 3. Experimental discharge voltage curves of 1st LIB strand for 0.2 C, 0.5 C, and 1 C.

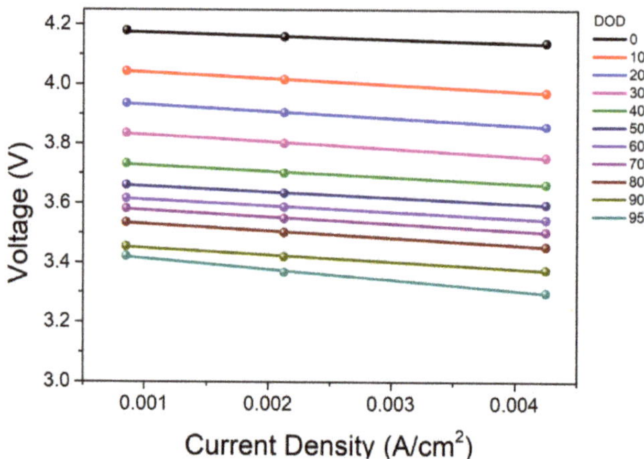

Figure 4. Strand voltage as a function of applied current density at various *DOD* levels during discharge.

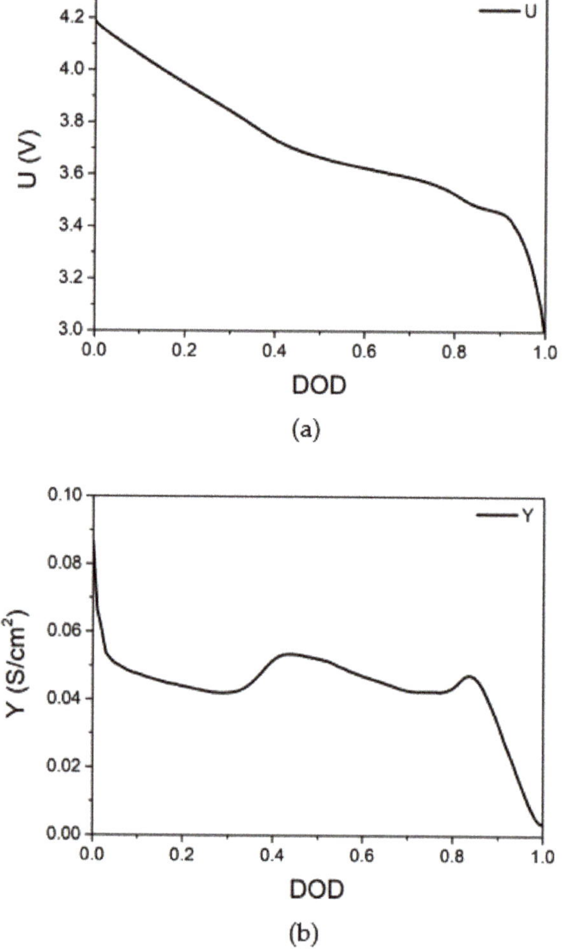

Figure 5. Fitting parameters of 1st LIB strand to calculate voltage behavior (**a**) *U* and (**b**) *Y*.

To verify the modeling methodology for the LIB strand, 0.2 C, 0.5 C, and 1 C discharge behaviors of the LIB strand were calculated using U and Y in Figure 5a,b. The experimental discharge voltage curves of the 1st LIB strand at 0.2, 0.5, and 1 C were compared with the modeling results, as shown in Figure 6. The comparison of the modeling results and experiments showed good agreement.

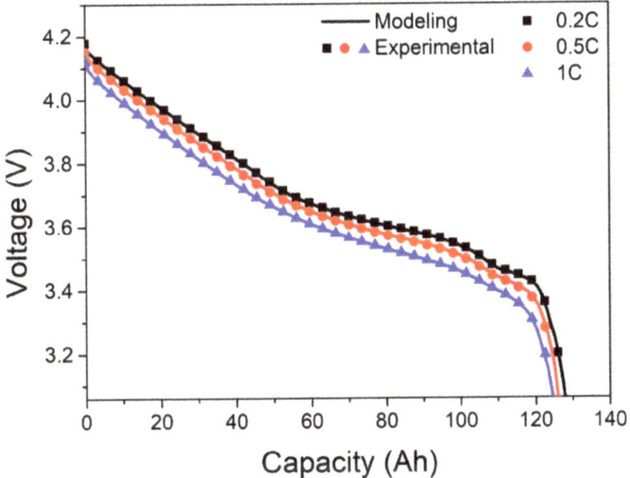

Figure 6. Comparison of discharge voltage behavior of modeling and experimental data at 0.2 C, 0.5 C, and 1 C of LIB strand.

4.2. Model Validation for LIB Module

It was found that the variation of U_n due to the change of the strand number, n, was negligible, if U_n is calculated as a function of DOD_n instead of DOD. Because the variation of Y_n was noticeable as the change of the strand number unlike U_n, the value of Y_n which is expressed in terms of β_n should be determined through the PSO algorithm along with Q_n. By using Equation (7) as the fitness function of the PSO algorithm, the Q_n and β_n that can render the best fit of the 0.2 C, 0.5 C, and 1 C voltage behaviors calculated from the modeling to the experimental data were obtained. Each LIB strand discharge voltage behavior was predicted by using the calculated Q_n and β_n values and it was compared with the experimental results. The modeling results are summarized in Table 1 in terms of the $RMSE$ values at different C-rates of the LIB strand and it was calculated by using the following formula:

$$RMSE = \sqrt{\sum_{i=1}^{T} \frac{\left(V_{exp} - V_{model}\right)^2}{T}} \qquad (8)$$

where T represents all sampling times in which the LIB strand was discharged with the discharge rates of 0.2 C, 0.5 C, and 1 C.

The results in Table 1 indicate that the model presented herein can accurately simulate the voltage behavior of the battery strand with an accuracy of 4 mV or less. The deviation in the capacity and internal resistance of the LIB cell is caused by the inconsistency in the electrochemical characteristics of the LIB cell that occurs during the process of material synthesis and cell manufacturing. Figure 7a–c show the capacity distributions at 0.2 C, 0.5 C, and 1 C. If we define the uniformity index (UI) of discharge capacity distribution as the difference between the maximum and minimum discharge capacities divided by the average value of discharge capacity, the UIs of 0.2, 0.5, and 1 C shown in Figure 7a–c are 0.006, 0.007, and 0.007, respectively. Although UI increases slightly with higher C rate, the values of UI at different C rates are very small and we may regard that the nonuniformity of the discharge capacity distribution is not significant. Figure 8 shows the distribution of β_n, which is a parameter related to the internal resistance of each LIB cell in the strand [29].

Table 1. RMSE of the modeling results and experimental data of discharge voltages of LIB strands at 0.2 C, 0.5 C, and 1 C.

Strand Number	RMSE (mV)		
	0.2 C	0.5 C	1 C
1	2.069	3.674	1.964
2	2.804	3.846	2.249
3	2.821	3.235	3.202
4	2.794	3.443	2.591
5	2.804	3.167	2.937
6	2.267	3.288	3.581
7	2.750	3.829	2.542
8	2.534	4.193	2.591
9	2.461	3.607	2.666
10	2.767	3.678	2.335
11	2.306	3.307	3.561
12	2.875	3.526	2.812
13	2.803	3.725	1.958
14	2.283	3.420	2.785

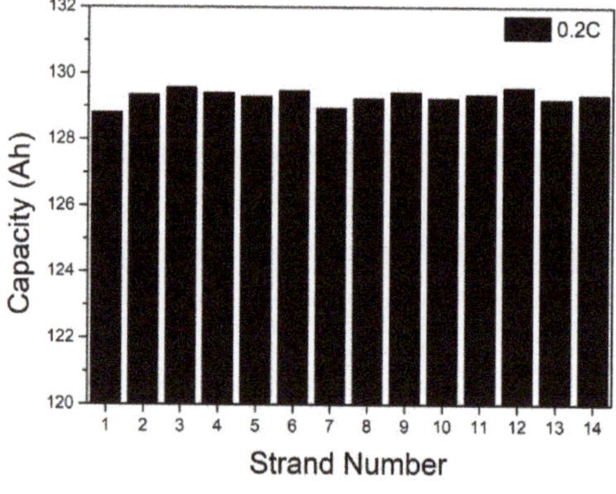

(**a**)

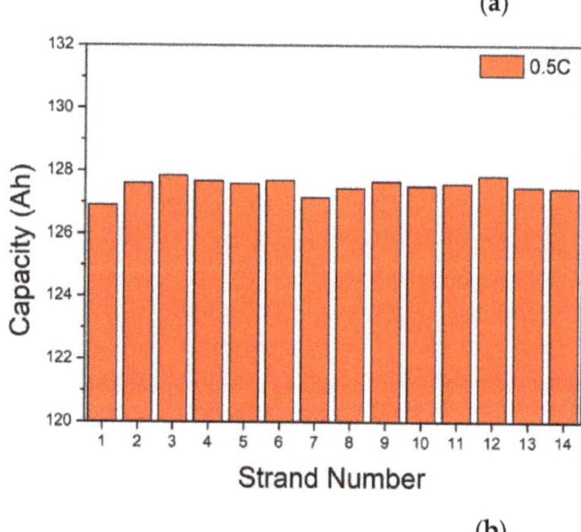

(**b**)

Figure 7. *Cont.*

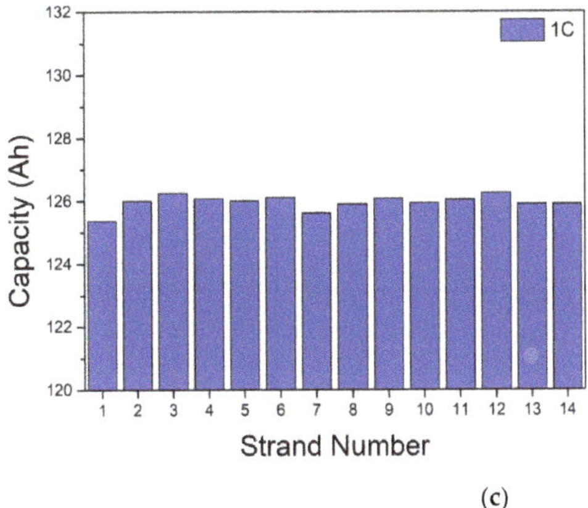

(**c**)

Figure 7. Modeling results of capacity distribution of LIB strands (**a**) 0.2 C, (**b**) 0.5 C, and (**c**) 1 C.

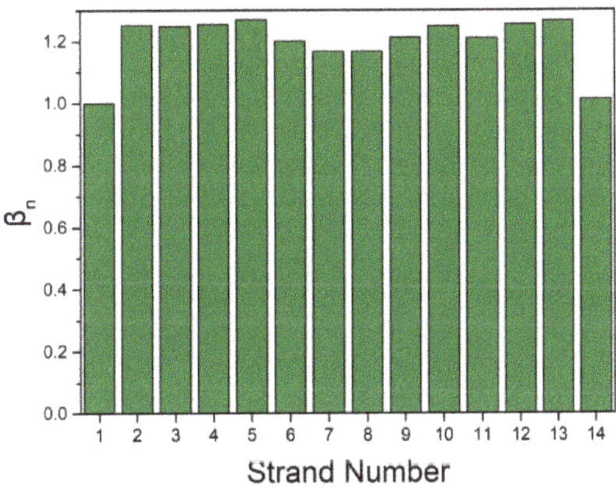

Figure 8. Modeling results of β_n distribution of LIB strands.

Based on the modeling methodology mentioned above, the discharge behaviors of the LIB module consisted of 14 strands connected in series were predicted. Figure 9 shows the comparison of the discharge voltage variations of the LIB module with time from modeling and those from experiment at the discharge rates of 0.2 C, 0.5 C, and 1 C. The voltage comparison results of the module were good; thus, the methodology presented in this paper may be justified.

4.3. Modeling of LIB Module Performance

When discharging in the LIB module, the termination voltage is terminated at 42 V and the cell termination voltage becomes 3.0 V considering the connection of 2P 14S. However, each cell maintains a different voltage level, and there are early discharge termination and overdischarge cells if we regard 3.0 V as the termination voltage. The module efficiency can be reduced in case of an early discharge. In the case of overdischarge, aging may be accelerated and cause serious safety problems. Therefore, predicting the end of voltage is important during the discharge of each cell. Figure 10 compares the end of voltage for each strand at 0.2, 0.5, and 1 C from the experimental with modeling results. The blue bars represent early discharge termination, where the termination voltage of the strand has not reached 3.0 V, and the red bars represent overdischarge, where the strand voltage has dropped below 3.0 V. The experimental results indicated by the black dots show a good fit with the modeling results.

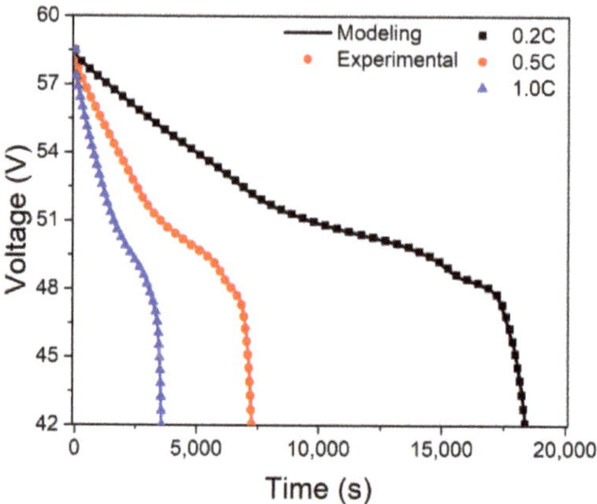

Figure 9. Comparison of modeling and experimental data of discharge behavior of LIB module at 0.2 C, 0.5 C, and 1 C.

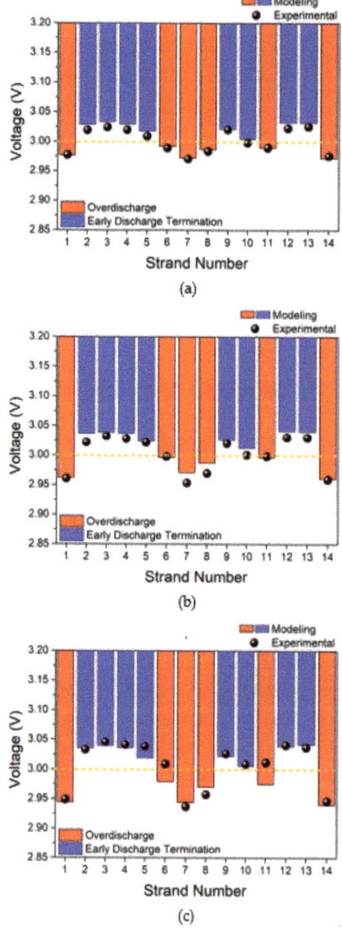

Figure 10. Comparison of experimental results with modeling results of the voltage of LIB strands at the end of discharge (**a**) 0.2 C, (**b**) 0.5 C, and (**c**) 1 C.

The voltage difference between the strands can be calculated based on the verified model, which accurately simulates the strand voltage in the LIB module. Figure 11 shows the difference between the maximum and minimum voltages measured by the sensor over time when discharging at 0.2, 0.5, and 1 C. The difference between the maximum and minimum voltages is defined as the voltage imbalance and it is compared with the modeling. Figure 11 shows that the voltage imbalance increases as the discharge current rate increases, and particularly at the end of the discharge, it rises very steeply. The comparison results of the model and experiment are in good agreement, indicating that the proposed modeling methodology is suitable for calculating the voltage deviation in the module.

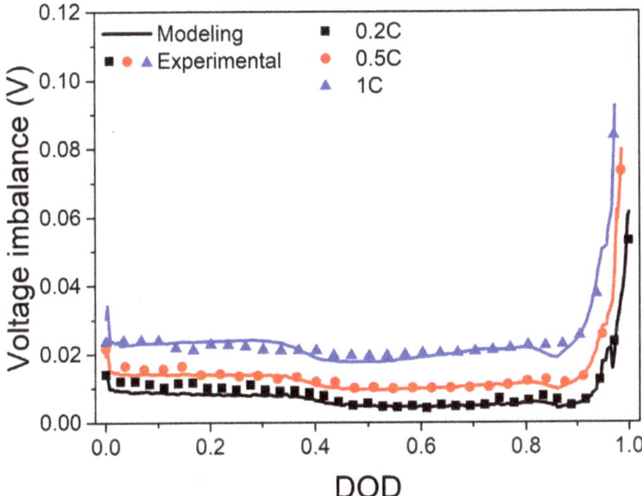

Figure 11. Comparison of modeling and experimental results of the difference between the maximum and minimum voltages between LIB strands according to *DOD*. The solid lines denote modeling results and symbols are the experimental data.

As proper cell balancing is not performed after charging, an initial *DOD* deviation occurs. The *DOD* range used by each strand during discharging is different. Figure 12a–c show the *DOD* range area from the initial *DOD* to the final *DOD* when discharging at 0.2 C, 0.5 C, and 1 C, respectively. The point at 3.0 V is indicated by a red dashed line. The initial *DOD* of each strand was obtained by interpolation into the OCV table, and the *DOD* use for each strand was calculated using Equation (6). In Figure 12a, when some strands fall below 3.0 V at 0.2 C discharge, an overdischarge occurs and *DOD* value exceeds 1.0. The rest of the strands are assessed as early termination discharge. In Figure 12b,c, the strand below the red dashed line is the early discharge termination, and the strand above the red dashed line is the overdischarge.

Variations in the voltage and *DOD* used in the LIB module are unavoidable owing to differences in the physical properties of the electrode materials and cell manufacturing process. Early discharge termination owing to these deviations reduces the efficiency of the LIB module. If overdischarge accumulates continuously, it can accelerate individual aging and cause serious damage to the battery system. Therefore, it is possible to accurately analyze and parametrize the cell deviation within the LIB module to calculate the difference by using the modeling methodology proposed in this study. Although the modeling approach was verified with data of only the discharge behavior, the deviation between the battery cells in a module or pack can be calculated in real time under various operating conditions using the modeling methodology presented herein. The modeling approach will be a good strategy for cell balancing and improving the energy efficiency in large-scale energy storage applications.

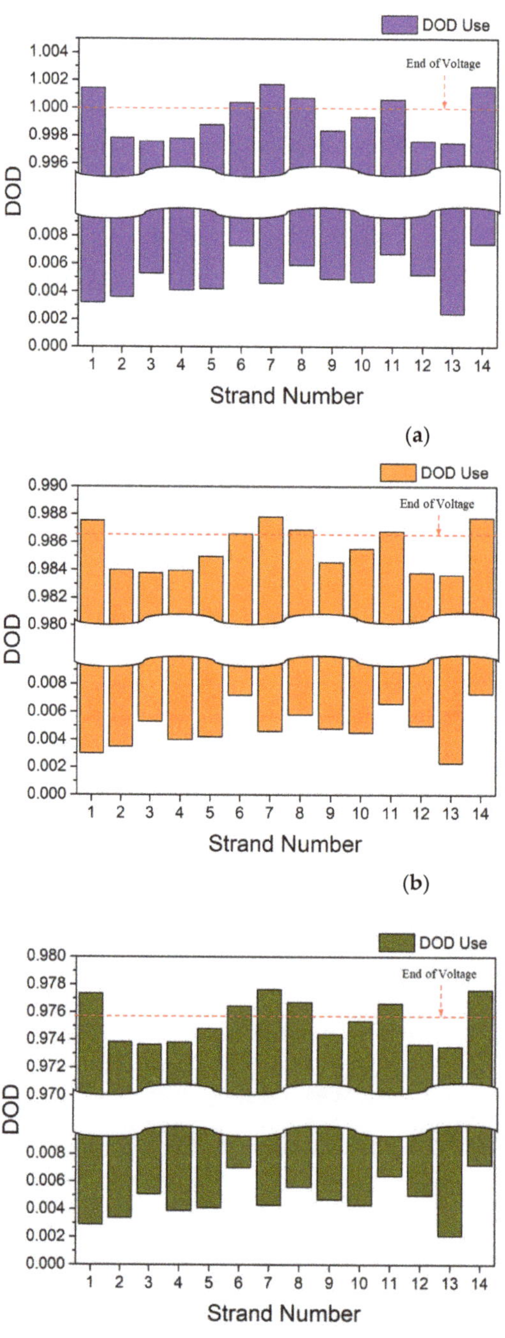

Figure 12. Modeling results of *DOD* use distribution of LIB strands during discharge (**a**) 0.2 C, (**b**) 0.5 C, and (**c**) 1 C.

5. Conclusions

We developed a mathematical model to accurately calculate the effect of cell variation during discharge in an LIB module. A simplified model that can simulate the behavior of an LIB strand, eliminating the computational burden, was applied. The discharge-voltage behavior of the LIB strand is modeled based on Ohm's law and charge conservation law using the simplified polarization characteristics of the electrode. Based on the modeling methodology validated on a single LIB strand, the capacity and internal resistance distributions of the 14 strands were calculated using the PSO algorithm. The calculation result was compared with each strand voltage measured by the sensor in the module and the calculation accuracy was expressed with RMSE. The modeling methodology in this study simulated the voltage behavior of a module composed of 2P 14S and showed an excellent agreement with the experimental data. Based on the verified modeling method, the voltage difference between the strands, discharge termination voltage, and *DOD* use distribution were modeled during discharge in the LIB module.

Author Contributions: D.L., S.K. and C.B.S. developed the methodology and performed modeling. All authors have read and agreed to the published version of the manuscript.

Funding: This research was supported by the Korea Institute of Energy Technology Evaluation and Planning (KETEP) grant funded by the Ministry of Trade, Industry & Energy (MOTIE) (No. 20206910100090) of the Republic of Korea. This study was partially funded by the Korea Electrotechnology Research Institute (KERI) Primary Research Program through the National Research Council of Science & Technology (NST), funded by the Ministry of Science and ICT (MSIT) (No. 22A01011), and the Technology Innovation Program (20011379, Development of advanced charge acceptance technology for charging power improvement of xEV battery system), funded by the Ministry of Trade, Industry & Energy (MOTIE, Korea).

Data Availability Statement: Data is contained within the article.

Conflicts of Interest: The authors declare no conflict of interest.

Nomenclature

Symbol	Description	Units
DOD	Depth of discharge	
DOD_i	DOD at $t = 0$	
DOD_n	DOD of nth LIB strand	
J	Current density between the electrodes	A cm^{-2}
n	LIB strand number	
Q_0	Nominal capacity per unit area	Ah cm^{-2}
Q_n	Q of LIB cell in nth LIB strand	Ah cm^{-2}
r_c	Resistance of cathode	Ω
r_a	Resistance of anode	Ω
t	Time	s
U	Polarization characteristic of the electrodes, intercept of the voltage-current curve	V
V_c	Voltage of cathode	V
V_a	Voltage of anode	V
V_{exp}	Experiment voltage	V
V_{model}	Modeling voltage	V
Y	Polarization characteristic of the electrodes, inverse of the slope of the voltage-current curve	S cm^{-2}
Y_n	Y of nth LIB strand	S cm^{-2}
Y_{ref}	Y of reference LIB strand	S cm^{-2}
β_n	Proportionality constant of internal resistance of nth strand	
Ω_a	Computational domain of anode	
Ω_c	Computational domain of cathode	

References

1. Kim, T.; Song, W.; Son, D.-Y.; Ono, L.K.; Qi, Y. Lithium-ion batteries: Outlook on present, future, and hybridized technologies. *J. Mater. Chem. A* **2019**, *7*, 2942–2964. [CrossRef]
2. Killer, M.; Farrokhseresht, M.; Paterakis, N.G. Implementation of large-scale Li-ion battery energy storage systems within the EMEA region. *Appl. Energy* **2020**, *260*, 114166. [CrossRef]
3. Omariba, Z.B.; Zhang, L.; Sun, D. Review of battery cell balancing methodologies for optimizing battery pack performance in electric vehicles. *IEEE Access* **2019**, *7*, 129335–129352. [CrossRef]
4. Naguib, M.; Kollmeyer, P.; Emadi, A. Lithium-ion battery pack robust state of charge estimation, cell inconsistency, and balancing: Review. *IEEE Access* **2021**, *9*, 50570–50582. [CrossRef]
5. Uzair, M.; Abbas, G.; Hosain, S. Characteristics of battery management systems of electric vehicles with consideration of the active and passive cell balancing process. *World Electr. Veh. J.* **2021**, *12*, 120. [CrossRef]
6. Feng, F.; Hu, X.; Liu, J.; Lin, X.; Liu, B. A review of equalization strategies for series battery packs: Variables, objectives, and algorithms. *Renew. Sustain. Energy Rev.* **2019**, *116*, 109464. [CrossRef]
7. Beck, D.; Dechent, P.; Junker, M.; Sauer, D.U.; Dubarry, M. Inhomogeneities and cell-to-cell variations in lithium-ion batteries, a review. *Energies* **2021**, *14*, 3276. [CrossRef]
8. Hemavathi, S. Overview of cell balancing methods for Li-ion battery technology. *Energy Storage* **2021**, *3*, 203.
9. Habib, A.K.M.A.; Hasan, M.K.; Mahmud, M.; Motakabber, S.M.A.; Ibrahimya, M.I.; Islam, S. A review: Energy storage system and balancing circuits for electric vehicle application. *IET Power Electron.* **2021**, *14*, 1–13. [CrossRef]
10. Xie, L.; Ren, D.; Wang, L.; Chen, Z.; Tian, G.; Amine, K.; He, X. A facile approach to high precision detection of cell-to-cell variation for Li-ion batteries. *Sci. Rep.* **2020**, *10*, 7182. [CrossRef]
11. Chang, F.; Roemer, F.; Baumann, M.; Lienkamp, M. Modelling and evaluation of battery packs with different numbers of paralleled cells. *World Electr. Veh. J.* **2018**, *9*, 8. [CrossRef]
12. Astaneh, M.; Andric, J.; Löfdahl, L.; Maggiolo, D.; Stopp, P.; Moghaddam, M.; Chapuis, M.; Ström, H. Calibration optimization methodology for lithium-ion battery pack model for electric vehicles in mining applications. *Energies* **2020**, *13*, 3532. [CrossRef]
13. Tran, N.-T.; Khan, A.B.; Nguyen, T.-T.; Kim, D.-W.; Choi, W. SOC Estimation of Multiple Lithium-Ion Battery Cells in a Module Using a Nonlinear State Observer and Online Parameter Estimation. *Energies* **2018**, *11*, 1620. [CrossRef]
14. Liu, R.; Zhang, C. An active balancing method based on SOC and capacitance for lithium-ion batteries in electric vehicles. *Front. Energy Res.* **2021**, *9*, 662. [CrossRef]
15. Lee, J.; Ahn, J.-H.; Lee, B.K. A novel Li-ion battery pack modeling considering single cell information and capacity variation. In Proceedings of the 2017 IEEE Energy Conversion Congress and Exposition (ECCE), Cincinnati, OH, USA, 1–5 October 2017; pp. 5242–5247.
16. Krupp, A.; Ferg, E.; Schuldt, F.; Derendorf, K.; Agert, C. Incremental capacity analysis as a state of health estimation method for lithium-ion battery modules with series-connected cells. *Batteries* **2021**, *7*, 2. [CrossRef]
17. Zilberman, I.; Ludwig, S.; Jossen, A. Cell-to-cell variation of calendar aging and reversible self-discharge in 18650 nickel-rich, silicon–graphite lithium-ion cells. *J. Energy Storage* **2019**, *26*, 100900. [CrossRef]
18. Hossain Lipu, M.S.; Hannan, M.A.; Karim, T.F.; Hussain, A.; Saad, M.H.M.; Ayob, A.; Miah, M.S.; Indra Mahlia, T.M. Intelligent algorithms and control strategies for battery management system in electric vehicles: Progress, challenges and future outlook. *J. Clean. Prod.* **2021**, *292*, 126044. [CrossRef]
19. Feng, F.; Hu, X.; Hu, L.; Hu, F.; Li, Y.; Zhang, L. Propagation mechanisms and diagnosis of parameter inconsistency within Li-ion battery packs. *Renew. Sustain. Energy Rev.* **2019**, *112*, 102–113. [CrossRef]
20. Wikipedia. Particle Swarm Optimization. Available online: https://en.wikipedia.org/wiki/Particle_swarm_optimization (accessed on 6 September 2022).
21. Kennedy, J.; Eberhart, R.C. Particle swarm optimization. In Proceedings of the IEEE International Conference on Neural Networks, Perth, Australia, 27 November 1995; pp. 1942–1948.
22. Kwon, K.H.; Shin, C.B.; Kang, T.H.; Kim, C. A two-dimensional modeling of a lithium-polymer battery. *J. Power Sources* **2006**, *163*, 151–157. [CrossRef]
23. Kim, U.S.; Shin, C.B.; Kim, C. Effect of electrode configuration on the thermal behavior of a lithium-polymer battery. *J. Power Sources* **2008**, *180*, 909–916. [CrossRef]
24. Kim, U.S.; Shin, C.B.; Kim, C. Modeling for the scale-up of a lithium-ion polymer battery. *J. Power Sources* **2009**, *189*, 841–846. [CrossRef]
25. Tiedemann, W.; Newman, J. Current and potential distribution in lead-acid battery plates. In *Battery Design and Optimization*; Gross, S., Ed.; The Electrochemical Society Inc.: Pennington, NJ, USA, 1979; pp. 39–49.
26. Newman, J.; Tiedemann, W. Potential and current distribution in electrochemical cells—Interpretation of the half-cell voltage measurements as a function of reference-electrode location. *J. Electrochem. Soc.* **1993**, *140*, 1961–1968. [CrossRef]
27. Gu, H. Mathematical analysis of a Zn / NiOOH cell. *J. Electrochem. Soc.* **1983**, *130*, 1459–1464. [CrossRef]
28. Kim, U.S.; Yi, J.; Shin, C.B.; Han, T.; Park, S. Modeling the dependence of the discharge behavior of a lithium-ion battery on the environmental temperature. *J. Electrochem. Soc.* **2011**, *158*, A611–A618. [CrossRef]
29. Yi, J.; Koo, B.; Shin, C.B.; Han, T.; Park, S. Modeling the effect of aging on the electrical and thermal behaviors of a lithium-ion battery during constant current charge and discharge cycling. *Comput. Chem. Eng.* **2017**, *99*, 31–39. [CrossRef]

30. Kim, U.S.; Yi, J.; Shin, C.B.; Han, T.; Park, S. Modelling the thermal behaviour of a lithium-ion battery during charge. *J. Power Sources* **2011**, *196*, 5115–5121. [CrossRef]
31. Kim, U.S.; Yi, J.; Shin, C.B.; Han, T.; Park, S. Modeling the thermal behaviors of a lithium-ion battery during constant-power discharge and charge operations. *J. Electrochem. Soc.* **2013**, *160*, A990–A995. [CrossRef]
32. Yi, J.; Kim, U.S.; Shin, C.B.; Han, T.; Park, S. Modeling the temperature dependence of the discharge behavior of a lithium-ion battery in low environmental temperature. *J. Power Sources* **2013**, *244*, 143–148. [CrossRef]
33. Yi, J.; Lee, J.; Shin, C.B.; Han, T.; Park, S. Modeling of the transient behaviors of a lithium-ion battery during dynamic cycling. *J. Power Sources* **2015**, *277*, 379–386. [CrossRef]
34. Koo, B.; Yi, J.; Lee, D.; Shin, C.B.; Han, T.; Park, S. Modeling the effect of fast charge scenario on the cycle life of a lithium-ion battery. *J. Electrochem. Soc.* **2018**, *165*, A3674–A3680. [CrossRef]
35. Lee, D.; Kim, B.; Shin, C.B. Modeling Fast Charge Protocols to Prevent Lithium Plating in a Lithium-Ion Battery. *J. Electrochem. Soc.* **2022**, *169*, 090502. [CrossRef]
36. Bruen, T.; Marco, J. Modelling and experimental evaluation of parallel connected lithium ion cells for an electric vehicle battery system. *J. Power Sources* **2016**, *310*, 91–101. [CrossRef]
37. Bo, Z.; Li, H.; Yang, H.; Li, C.; Wu, S.; Xu, C.; Xiong, G.; Mariotti, D.; Yan, J.; Cen, K.; et al. Combinatorial atomistic-to-AI prediction and experimental validation of heating effects in 350 F supercapacitor modules. *Int. J. Heat Mass Transf.* **2021**, *171*, 121075. [CrossRef]

MDPI

St. Alban-Anlage 66

4052 Basel

Switzerland

Tel. +41 61 683 77 34

Fax +41 61 302 89 18

www.mdpi.com

Energies Editorial Office

E-mail: energies@mdpi.com

www.mdpi.com/journal/energies